Machine Learning: Foundations, Methodologies, and Applications

Series Editors

Kay Chen Tan, Department of Computing, Hong Kong Polytechnic University, Hong Kong, China

Dacheng Tao, University of Technology, Sydney, Australia

Books published in this series focus on the theory and computational foundations, advanced methodologies and practical applications of machine learning, ideally combining mathematically rigorous treatments of a contemporary topics in machine learning with specific illustrations in relevant algorithm designs and demonstrations in real-world applications. The intended readership includes research students and researchers in computer science, computer engineering, electrical engineering, data science, and related areas seeking a convenient medium to track the progresses made in the foundations, methodologies, and applications of machine learning.

Topics considered include all areas of machine learning, including but not limited to:

- Decision tree
- Artificial neural networks
- Kernel learning
- Bayesian learning
- Ensemble methods
- Dimension reduction and metric learning
- Reinforcement learning
- Meta learning and learning to learn
- Imitation learning
- Computational learning theory
- Probabilistic graphical models
- Transfer learning
- Multi-view and multi-task learning
- Graph neural networks
- Generative adversarial networks
- Federated learning

This series includes monographs, introductory and advanced textbooks, and state-of-the-art collections. Furthermore, it supports Open Access publication mode.

More information about this series at https://link.springer.com/bookseries/16715

Alexander Jung

Machine Learning

The Basics

 Springer

Alexander Jung
Department of Computer Science
Aalto University
Espoo, Finland

ISSN 2730-9908   ISSN 2730-9916   (electronic)
Machine Learning: Foundations, Methodologies, and Applications
ISBN 978-981-16-8195-0   ISBN 978-981-16-8193-6   (eBook)
https://doi.org/10.1007/978-981-16-8193-6

This Springer imprint is published by the registered company Springer Nature Singapore Pte Ltd.
The registered company address is: 152 Beach Road, #21-01/04 Gateway East, Singapore 189721,
Singapore

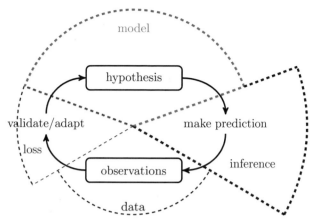

Fig. 1 Machine learning combines three main components: data, model and loss. Machine learning methods implement the scientific principle of "trial and error". These methods continuously validate and refine a model based on the loss incurred by its predictions about a phenomenon that generates data.

Preface

Machine learning (ML) influences our daily lives in several aspects. We routinely ask ML empowered smartphones to suggest lovely restaurants or to guide us through a strange place. ML methods have also become standard tools in many fields of science and engineering. ML applications transform human lives at unprecedented pace and scale.

This book portrays ML as the combination of three basic components: data, model and loss. ML methods combine these three components within computationally efficient implementations of the basic scientific principle "trial and error". This principle consists of the continuous adaptation of a hypothesis about a phenomenon that generates data.

ML methods use a hypothesis map to compute predictions of a quantity of interest (or higher level fact) that is referred to as the label of a data point. A hypothesis map reads in low level properties (referred to as features) of a data point and delivers the prediction for the label of that data point. ML methods choose or learn a hypothesis map from a (typically very) large set of candidate maps. We refer to this set as of candidate maps as the hypospace or model underlying an ML method.

The adaptation or improvement of the hypothesis is based on the discrepancy between predictions and observed data. ML methods use a loss function to quantify this discrepancy.

A plethora of different ML methods is obtained by combining different design choices for the data representation, model and loss. ML methods also differ vastly in their practical implementations which might obscure their unifying basic principles.

Deep learning methods use cloud computing frameworks to train large models on large datasets. Operating on a much finer granularity for data and computation, linear (least squares) regression can be implemented on small embedded systems. Nevertheless, deep learning methods and linear regression use the same principle of iteratively updating a model based on the discrepancy between model predictions and actual observed data.

We believe that thinking about ML as combinations of three components given by data, model and lossfunc helps to navigate the steadily growing offer for ready-to-use

ML methods. Our three-component picture allows a unified treatment of ML techniques, such as early stopping, privacy-preserving ML and xml, that seem quite unrelated at first sight. For example, the regularization effect of the early stopping technique in gradient-based methods is due to the shrinking of the effective hypospace. Privacy-preserving ML methods can be obtained by particular choices for the features used to characterize data points (see Sect. 9.5). Explainable ML methods can be obtained by particular choices for the hypospace and lossfunc (see Chap. 10).

To make good use of ML tools it is instrumental to understand its underlying principles at the appropriate level of detail. It is typically not necessary to understand the mathematical details of advanced optimization methods to successfully apply deep learning methods. On a lower level, this tutorial helps ML engineers choose suitable methods for the application at hand. The book also offers a higher level view on the implementation of ML methods which is typically required to manage a team of ML engineers and data scientists.

Espoo, Finland   Alexander Jung

Acknowledgements

This book grew from lecture notes prepared for the courses CS-E3210 "Machine Learning: Basic Principles", CS-E4800 "Artificial Intelligence", CS-EJ3211 "Machine Learning with Python", CS-EJ3311 "Deep Learning with Python" and CS-C3240 "Machine Learning" offered at Aalto University and within the Finnish university network fitech.io. This tutorial is accompanied by practical implementations of ML methods in MATLAB and Python https://github.com/alexjungaalto/.

This text benefited from the numerous feedback of the students within the courses that have been (co-)taught by the author. The author is indebted to Shamsiiat Abdurakhmanova, Tomi Janhunen, Yu Tian, Natalia Vesselinova, Linli Zhang, Ekaterina Voskoboinik, Buse Atli, Stefan Mojsilovic for carefully reviewing early drafts of this tutorial. Some of the figures have been generated with the help of Linli Zhang. The author is grateful for the feedback received from Jukka Suomela, Väinö Mehtola, Oleg Vlasovetc, Anni Niskanen, Georgios Karakasidis, Joni Pääkkö, Harri Wallenius and Satu Korhonen.

Contents

1	**Introduction**		1
	1.1	Relation to Other Fields	4
		1.1.1 Linear Algebra	5
		1.1.2 Optimization	6
		1.1.3 Theoretical Computer Science	6
		1.1.4 Information Theory	7
		1.1.5 Probability Theory and Statistics	9
		1.1.6 Artificial Intelligence	10
	1.2	Flavours of Machine Learning	12
		1.2.1 Supervised Learning	12
		1.2.2 Unsupervised Learning	13
		1.2.3 Reinforcement Learning	14
	1.3	Organization of this Book	15
	References		17
2	**Components of ML**		19
	2.1	The Data	19
		2.1.1 Features	21
		2.1.2 Labels	26
		2.1.3 Scatterplot	28
		2.1.4 Probabilistic Models for Data	28
	2.2	The Model	30
		2.2.1 Parametrized Hypothesis spaces	32
		2.2.2 The Size of a Hypothesis Space	35
	2.3	The Loss	37
		2.3.1 Loss Functions for Numeric Labels	39
		2.3.2 Loss Functions for Categorical Labels	40
		2.3.3 Loss Functions for Ordinal Label Values	43
		2.3.4 Empirical Risk	44
		2.3.5 Regret	47
		2.3.6 Rewards as Partial Feedback	48

2.4   Putting Together the Pieces   48
    2.5   Exercises   50
    References   55

3   **The Landscape of ML**   57
    3.1   Linear Regression   57
    3.2   Polynomial Regression   59
    3.3   Least Absolute Deviation Regression   60
    3.4   The Lasso   61
    3.5   Gaussian Basis Regression   62
    3.6   Logistic Regression   64
    3.7   Support Vector Machines   66
    3.8   Bayes Classifier   68
    3.9   Kernel Methods   69
    3.10   Decision Trees   70
    3.11   Deep Learning   72
    3.12   Maximum Likelihood   74
    3.13   Nearest Neighbour Methods   75
    3.14   Deep Reinforcement Learning   76
    3.15   LinUCB   77
    3.16   Exercises   78
    References   79

4   **Empirical Risk Minimization**   81
    4.1   The Basic Idea of Empirical Risk Minimization   83
    4.2   Computational and Statistical Aspects of ERM   84
    4.3   ERM for Linear Regression   86
    4.4   ERM for Decision Trees   89
    4.5   ERM for Bayes Classifiers   91
    4.6   Training and Inference Periods   94
    4.7   Online Learning   94
    4.8   Exercise   96
    References   97

5   **Gradient-Based Learning**   99
    5.1   The GD Step   100
    5.2   Choosing Step Size   102
    5.3   When to Stop?   103
    5.4   GD for Linear Regression   103
    5.5   GD for Logistic Regression   106
    5.6   Data Normalization   107
    5.7   Stochastic GD   108
    5.8   Exercises   111
    References   112

6 Model Validation and Selection 113
  6.1 Overfitting 115
  6.2 Validation 117
    6.2.1 The Size of the Validation Set 119
    6.2.2 k-Fold Cross Validation 120
    6.2.3 Imbalanced Data 120
  6.3 Model Selection 122
  6.4 A Probabilistic Analysis of Generalization 126
  6.5 The Bootstrap 130
  6.6 Diagnosing ML 131
  6.7 Exercises 133
  References 134

7 Regularization 135
  7.1 Structural Risk Minimization 137
  7.2 Robustness 140
  7.3 Data Augmentation 141
  7.4 Statistical and Computational Aspects of Regularization 144
  7.5 Semi-Supervised Learning 147
  7.6 Multitask Learning 148
  7.7 Transfer Learning 149
  7.8 Exercises 150
  References 151

8 Clustering 153
  8.1 Hard Clustering with k-Means 155
  8.2 Soft Clustering with Gaussian Mixture Models 162
  8.3 Connectivity-Based Clustering 167
  8.4 Clustering as Preprocessing 169
  8.5 Exercises 170
  References 170

9 Feature Learning 173
  9.1 Basic Principle of Dimensionality Reduction 174
  9.2 Principal Component Analysis 176
    9.2.1 Combining PCA with Linear Regression 178
    9.2.2 How to Choose Number of PC? 179
    9.2.3 Data Visualisation 179
    9.2.4 Extensions of PCA 179
  9.3 Feature Learning for Non-numeric Data 181
  9.4 Feature Learning for Labeled Data 183
  9.5 Privacy-Preserving Feature Learning 185
  9.6 Random Projections 186
  9.7 Dimensionality Increase 187
  9.8 Exercises 187
  References 188

10   Transparent and Explainable ML 189

　　10.1   A Model Agnostic Method 191

　　　　　10.1.1   Probabilistic Data Model for XML 193

　　　　　10.1.2   Computing Optimal Explanations 194

　　10.2   Explainable Empirical Risk Minimization 196

　　10.3   Exercises 199

　　References 199

Glossary 201

Index 211

Symbols

Sets

$a := b$	This statement defines a to be shorthand for b.
$\mathbb{N}$	The set of natural numbers 1, 2, ...
$\mathbb{R}$	The set of real numbers x [2].
$\mathbb{R}_+$	The set of non-negative real numbers $x \geq 0$.
$\{0, 1\}$	The set consisting of two real number 0 and 1.
$[0, 1]$	The closed interval of real numbers x with $0 \leq x \leq 1$.

Matrices and Vectors

$\mathbf{I}$	The identity matrix having diagonal entries equal to one and every off diagonal entry equal to zero.
$\mathbb{R}^n$	The set of vectors that consist of n real-valued entries.
$\mathbf{x} = (x_1, \ldots, x_n)^T$	A vector of length n. The jth entry of the vector is denoted as x_j.
$\|\mathbf{x}\|_2$	The Euclidean (or "ℓ_2") norm of the vector $\mathbf{x} = (x_1, \ldots, x_n)^T$ given as $\|\mathbf{x}\|_2 := \sqrt{\sum_{j=1}^n x_j^2}$.
$\|\mathbf{x}\|$	Some norm of the vector $\mathbf{x}$ [1]. Unless specified otherwise, we mean the Euclidean norm $\|\mathbf{x}\|_2$.
$\mathbf{x}^T$	The transpose of a vector $\mathbf{x}$ that is considered as a single column matrix. The transpose can be interpreted as a single-row matrix $(x_1, \ldots, x_n)$.
$\mathbf{A}^T$	The transpose of a matrix $\mathbf{A}$. A square matrix is called symmetric if $\mathbf{A} = \mathbf{A}^T$
$\mathbb{S}_+^n$	The set of all (psd) $n \times n$ matrices.

Machine Learning

i	A generic index $i = 1, 2, \ldots$, used to enumerate the data points within a dataset.
m	The number of data points in (the size of) a dataset.
n	The number of individual features used to characterize a data point.
x_j	The jth individual feature of a data point.
$\mathbf{x}$	The feature vector $\mathbf{x} = (x_1, \ldots, x_n)^T$ of a data point whose entries are the individual features of the data point.
$\mathbf{z}$	Beside the symbol x, we sometimes use as another symbol to denote a vector whose entries are features of a data point. We need two different symbols to denote feature vectors for the discussion feature learning methods in Chap. 9.
$\mathbf{x}^{(i)}$	The feature vector of the ith data point within a dataset.
$x_j^{(i)}$	The jth feature of the ith data point within a dataset.
y	The label (quantity of interest) of a data point.
$y^{(i)}$	The label of the ith data point.
$(\mathbf{x}^{(i)}, y^{(i)})$	The features and the label of the ith data point within a dataset.
$h(\cdot)$	A hypothesis map that reads in the features $\mathbf{x}$ of a data point and outputs the predicted label $\hat{y} = h(\mathbf{x})$.
x_j	The j-th feature of a data point. The first feature of a given data point is denoted as x_1, the second feature x_2 and so on.
$L((\mathbf{x}, y), h)$	The loss incurred by predicting the label y of a data point with feature vector $\mathbf{x}$ using the value $\hat{y} = h(\mathbf{x})$ obtained from evaluating the hypothesis $h \in \mathcal{H}$ at the feature vector $\mathbf{x}$.
E_v	The validation error of a hypothesis, which is the average loss computed on a validation set.
$\widehat{L}(h\|\mathcal{D})$	The emprisk or average loss incurred by the predictions of hypothesis h for the data points in the dataset $\mathcal{D}$.
E_t	The trainer of a hypothesis h, which is the average loss incurred by h on labeled data points that form a training set.
t	A discrete-time index $t = 0, 1, \ldots$ used to enumerate a sequence to temporal events (time instants).
t	A generic index used to enumerate a finite set of learning tasks within a multi-task learning problem (see Sect. 7.6).
λ	A regularization parameter that is used to scale the regularization term that is added to the empirical risk in structural risk minimization (SRM).
$\lambda_j(\mathbf{Q})$	The jth eigenvalue (sorted either ascending or descending) of a psd matrix $\mathbf{Q}$. We also use the shorthand λ_j if the corresponding matrix is clear from context.
$f(\cdot)$	The activation function used by an artificial neuron within an artificial neural network (ANN).

References

1. G.H. Golub, C. F. Van Loan. Matrix Computations. (Johns Hopkins University Press, Baltimore, MD, 3rd edition, 1996)
2. W. Rudin. Real and Complex Analysis. (McGraw-Hill, New York, 3rd edition, 1987)

Chapter 1
Introduction

Consider waking up one winter morning in Finland and looking outside the window (see Fig. 1.1). It seems to become a nice sunny day which is ideal for a ski trip. To choose the right gear (clothing, wax) it is vital to have some idea for the maximum daytime temperature which is typically reached around early afternoon. If we expect a maximum daytime temperature of around plus 5 degrees, we might not put on the extra warm jacket but rather take only some extra shirt for change.

We can use ML to learn a predictor for the maximum daytime temperature for the specific day depicted in Fig. 1.1. The prediction shall be based solely on the minimum temperature observed in the morning of that day. ML methods can learn a predictor in a data-driven fashion using historic weather observations provided by the Finnish Meteorological Institute. We can download the recordings of minimum and maximum daytime temperature for the most recent days and denote the resulting dataset by

$$\mathcal{D} = \left\{ \mathbf{z}^{(1)}, \ldots, \mathbf{z}^{(m)} \right\}. \tag{1.1}$$

Each data point $\mathbf{z}^{(i)} = \left(x^{(i)}, y^{(i)} \right)$, for $i = 1, \ldots, m$, represents some previous day for which the minimum and maximum daytime temperature $x^{(i)}$ and $y^{(i)}$ has been recorded. We depict the data (1.1) in Fig. 1.2. Each dot in Fig. 1.2 represents a specific day with minimum temperature x and maximum temperature y.

ML methods learn a hypothesis $h(x)$, that reads in the minimum temperature x and delivers a prediction (forecast or approximation) $\hat{y} = h(x)$ for the maximum daytime temperature y. Every practical ML method uses a particular hypothesis space out of which the hypothesis h is chosen. This hypothesis space of candidates for the hypothesis map is an important design choice and might be based on domain knowledge.

In what follows, we illustrate how to use domain knowledge to motivate a choice for the hypothesis space. Let us assume that the minimum and maximum daytime temperature of an arbitrary day are approximately related via

© The Author(s), under exclusive license to Springer Nature Singapore Pte Ltd. 2022
A. Jung, *Machine Learning*, Machine Learning: Foundations, Methodologies,
and Applications, https://doi.org/10.1007/978-981-16-8193-6_1

Fig. 1.1 Looking outside the window during the morning of a winter day in Finland

Fig. 1.2 Each dot represents a specific day that is characterized by its minimum daytime temperature x as feature and its maximum daytime temperature y as label. These temperatures are measured at some Finnish Meteorological Institute weather station

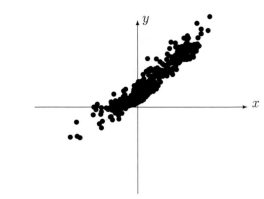

$$y \approx w_1 x + w_0 \text{ with some weights } w_1 \in \mathbb{R}_+, w_0 \in \mathbb{R}. \qquad (1.2)$$

The assumption (1.2) reflects the intuition (domain knowledge) that the maximum daytime temperature y should be higher for days with a higher minimum daytime temperature x. The assumption (1.2) contains two weights w_1 and w_0. These weights are tuning parameters that allow for some flexibility in our assumption. We require the weight w_1 to be non-negative but otherwise leave these weights unspecified for the time being. The main subject of this book are ML methods that can be used to learn suitable values for the weights w_1 and w_0 in a data-driven fashion.

Before we detail how ML can be used to find or learn good values for the weights w_0 in w_1 in (1.2) let us interpret them. The weight w_1 in (1.2) can be interpreted as the relative increase in the maximum daytime temperature for an increased minimum daytime temperature. Consider an earlier day with recorded maximum daytime temperature of 10 degrees and minimum daytime temperature of 0 degrees. The assumption (1.2) then means that the maximum daytime temperature for another other day with minimum daytime temperature of $+1$ degrees would be $10 + w_1$ degrees. The second weight w_0 in our assumption (1.2) can be interpreted as the maximum daytime temperature that we anticipate for a day with minimum daytime temperature equal to 0.

Fig. 1.3 Three hypothesis maps of the form (1.3)

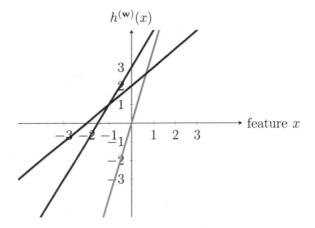

Given the assumption (1.2), it seems reasonable to restrict the ML method to only consider linear maps

$$h(x) := w_1 x + w_0 \text{ with some weights } w_1 \in \mathbb{R}_+, w_0 \in \mathbb{R}. \tag{1.3}$$

Since we require $w_1 \geq 0$, the map (1.3) is monotonically increasing with respect to the argument x. Therefore, the prediction $h(x)$ for the maximum daytime temperature becomes higher with higher minimum daytime temperature x.

The expression (1.3) defines a whole ensemble of hypothesis maps. Each individual map corresponding to a particular choice for $w_1 \geq 0$ and w_0. We refer to such an ensemble of potential predictor maps as the model or hypothesis space that is used by a ML method.

We say that the map (1.3) is parameterized by the weight vector $\mathbf{w} = (w_1, w_0)$ and indicate this by writing $h^{(\mathbf{w})}$. For a given weight vector $\mathbf{w} = (w_1, w_0)^T$, we obtain the map $h^{(\mathbf{w})}(x) = w_1 x + w_0$. Figure 1.3 depicts three maps $h^{(\mathbf{w})}$ obtained for three different choices for the weights $\mathbf{w}$.

ML would be trivial if there is only one single hypothesis. Having only a single hypothesis means that there is no need to try out different hypotheses to find the best one. To enable ML, we need to choose between a whole space of different hypotheses. ML methods are computationally efficient methods to choose (learn) a good hypothesis out of (typically very large) hypothesis spaces. The hypothesis space constituted by the maps (1.3) for different weights is uncountably infinite.

To find, or **learn**, a good hypothesis out of the infinite set (1.3), we need to somehow assess the quality of a particular hypothesis map. ML methods use a loss function for this purpose. A loss function is used to quantify the difference between the actual data and the predictions obtained from a hypothesis map (see Fig. 1.4). One widely-used example of a loss function is the squared error loss $(y - h(x))^2$. Using this loss function, ML methods learn a hypothesis map out of the model (1.3) by tuning w_1, w_0 to minimize the average loss

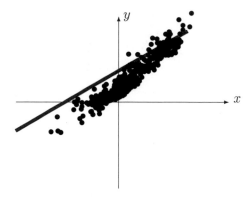

Fig. 1.4 Each dot represents a specific days that is characterized by its minimum daytime temperature x and its maximum daytime temperature y. We also depict a straight line representing a linear predictor map. A main principle of ML methods is to learn a predictor (or hypothesis) map with minimum discrepancy between predictor map and data points. Different ML methods use different types of predictor maps (hypothesis space) and loss functions to quantify the discrepancy between hypothesis and actual data points

$$(1/m) \sum_{i=1}^{m} \left(y^{(i)} - h\left(x^{(i)}\right)\right)^2.$$

The above weather prediction is prototypical for many other ML applications. Figure 1.4 illustrates the typical workflow of a ML method. Starting from some initial guess, ML methods repeatedly improve their current hypothesis based on (new) observed data.

Using the current hypothesis, ML methods make predictions or forecasts about future observations. The discrepancy between the predictions and the actual observations, as measured using some loss function, is used to improve the hypothesis. Learning happens during improving the current hypothesis based on the discrepancy between its predictions and the actual observations.

ML methods must start with some initial guess or choice for a good hypothesis. This initial guess can be based on some prior knowledge or domain expertise [1]. While the initial guess for a hypothesis might not be made explicit in some ML methods, each method must use such an initial guess. In our weather prediction application discussed above, we used the linear model (1.2) as the initial hypothesis.

1.1   Relation to Other Fields

ML builds on concepts from several other scientific fields. Conversely, ML provides important tools for many other scientific fields.

1.1.1 Linear Algebra

Modern ML methods are computationally efficient methods to fit high-dimensional models to large amounts of data. The models underlying state-of-the-art ML methods can contain billions of tunable or learnable parameters. To make ML methods computationally efficient we need to use suitable representations for data and models.

Maybe the most widely used mathematical structure to represent data is the Euclidean space $\mathbb{R}^n$ with some dimension $n \in \mathbb{N}$ [2]. The rich algebraic and geometric structure of $\mathbb{R}^n$ allows us to design of ML algorithms that can process vast amounts of data to quickly update a model (parameters). Figure 1.5 depicts the Euclidean space $\mathbb{R}^n$ for $n = 2$, which is used to construct scatterplots.

The scatterplot in Fig. 1.2 depicts data points (representing individual days) as vectors in the Euclidean space $\mathbb{R}^2$. For a given data point, we obtain its associated vector $\mathbf{z} = (x, y)^T$ in $\mathbb{R}^2$ by stacking the minimum daytime temperature x and the maximum daytime temperature y into the vector $\mathbf{z}$ of length two.

We can use the Euclidean space $\mathbb{R}^n$ not only to represent data points but also to represent models for these data points. One such class of models is obtained by linear maps on $\mathbb{R}^n$. Figure 1.3 depicts some examples for such linear maps. We can then use the geometric structure of $\mathbb{R}^n$, defined by the Euclidean norm, to search for the best model. As an example, we could search for the linear model, represented by a straight line, such that the average (Euclidean) distance to the data points in Fig. 1.2 is as small as possible (see Fig. 1.4). The properties of linear structures are studied within linear algebra [3]. Some important ML methods, such as linear classifier (see Sect. 3.1) or principal component analysis (see Sect. 9.2) are direct applications of methods from linear algebra.

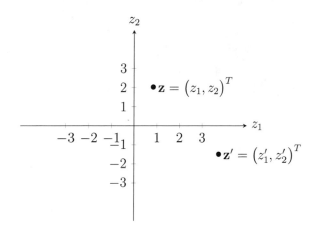

Fig. 1.5 The Euclidean space $\mathbb{R}^2$ is constituted by all vectors (or points) $\mathbf{z} = (z_1, z_2)^T$ (with $z_1, z_2 \in \mathbb{R}$) together with the inner product $\mathbf{z}^T \mathbf{z}' = z_1 z_1' + z_2 z_2'$

1.1.2 Optimization

A main design principle for ML methods is the formulation of ML problems as optimization problems [4]. The weather prediction problem above can be formulated as the problem of optimizing (minimizing) the prediction error for the maximum daytime temperature. Many ML methods are obtained by straightforward applications of optimization methods to the optimization problem arising from a ML problem (or application).

The statistical and computational properties of such ML methods can be studied using tools from the theory of optimization. What sets the optimization problems in ML apart from "plain vanilla" optimization problems (see Fig. 1.6a) is that we rarely have perfect access to the objective function to be minimized. ML methods learn a hypothesis by minimizing a noisy or even incomplete version (see Fig. 1.6b) of the actual objective which is defined using an expectation over an unknown probability distribution. Section 4 discusses methods that are based on estimating the objective function by empirical averages that are computed over a set of data points (forming a training set).

1.1.3 Theoretical Computer Science

Practical ML methods form a specific subclass of computing systems. Indeed, ML methods apply a sequence of computational operations to input data. The result of these computational operations are the predictions delivered to the user of the ML method. The interpretation of ML as computational systems allows to use tools from theoretical computer science to study the feasibility and intrinsic difficulty of ML

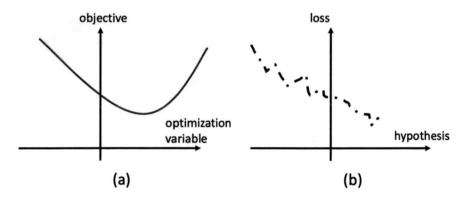

Fig. 1.6 a A simple optimization problem consists of finding the values of an optimization variable that results in the minimum objective value. **b** ML methods learn (find) a hypothesis by minimizing a loss that is a noisy and incomplete version of the actual objective

problems. Even if a ML problem can be solved in theoretical sense, every practical ML method must fit the available computational infrastructure [5, 6].

The available computational resources, such as processor time, memory and communication bandwidth, can vary significantly between different infrastructures. One example for such a computational infrastructure is a single desktop computer. Another example for a computational infrastructure is a cloud computing service which distributes data and computation over large networks of physical computers [7].

The focus of this book is on ML methods that can be understood as numerical optimization algorithms (see Chaps. 4 and 5). Most of these ML methods amount to (a large number of) matrix operations such as matrix multiplication or matrix inversion [8]. Numerical linear algebra provides a vast algorithmic toolbox for the design of such ML methods [3, 9]. The recent success of ML methods in several application domains might be attributed to their efficient use of matrices to represent data and models. Using this representation allows us to implement the resulting ML methods using highly efficient hard- and software implementations for numerical linear algebra [10].

1.1.4 Information Theory

Information theory studies the problem of communication via noisy channels [11–14]. Figure 1.7 depicts the most simple communication problem that consists of an information source that wishes communicate a message m over an imperfect (or noisy) channel to a receiver. The receiver tries to reconstruct (or learn) the original message based solely on the noisy channel output. Two main goals of information theory are (i) the characterization of conditions that allow reliable, i.e., nearly error-free, communication and (ii) the design of efficient transmitter (coding and modulation) and receiver (demodulation and decoding) methods.

It turns out that many concepts from information theory are very useful for the analysis and design of ML methods. As a point in case, Chap. 10 discusses the application of information-theoretic concepts to the design of explainable machine learning methods. On a more fundamental level, we can identify two core communication problems that arise within ML. These communication problems correspond, respectively, to the inference (making a prediction) and the learning (adjusting or improving the current hypothesis) step of a ML method (see Fig.1.4).

We can an interpret the inference step of ML as the problem of decoding the true label of a data point for which we only know its features. This communication problem is depicted in Fig. 1.7b. Here the message to be communicated is the true label of a random data point. This data point is "communicated" over a channel that only passes through its features. The inference step within a ML method then tries to decode the original message (true label) from the channel output (features) resulting in the predicted label. A recent line of work used this communication problem to study deep learning methods [11].

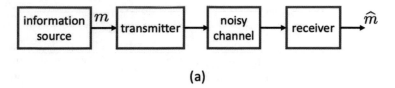

(a)

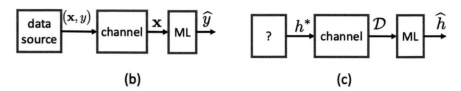

(b)   (c)

Fig. 1.7 **a** A basic communication system involves an information source that emits a message m. The message is processed by some transmitter and passed through a noisy channel. The receiver tries to recover the original message m by computing the decoded message $\hat{m}$. **b** The inference step of ML (see Fig. 1.4) corresponds to a communication problem with an information source emitting a data point with features $\mathbf{x}$ and label y. The ML method receives the features $\mathbf{x}$ and, in an effort to recover the true label y, computes the predicted label $\hat{y}$. **c** The learning or adaptation step of ML (see Fig.1.4) solves a communication problem with some source that selects a true (but unknown) hypothesis h^* as the message. The message is passed through an abstract channel that outputs a set $\mathcal{D}$ of labeled data points which are used as the training set by an ML method. The ML method tries to decode the true hypothesis resulting in the learnt the hypothesis $\hat{h}$

A second core communication problem of ML corresponds to the problem of learning (or adjusting) a hypothesis (see Fig. 1.7c). In this problem, the source selects some "true" hypothesis as message. This message is then communicated to an abstract channel that models the data generation process. The output of this abstract channel are data points in a training set $\mathcal{D}$ (see Chap. 4). The learning step of a ML method, such as empirical risk minimization of Chap. 4, then amounts to the decoding of the message (true hypothesis) based on the channel output (training set). There is significant line or research that uses the communication problem in Fig. 1.7c to characterize the fundamental limits of ML problems and methods such as the minimum required number of training data points that makes learning feasible [15–19].

The relevance of information theoretic concepts and methods for ML is boosted by the recent trend towards distributed or federated ML [20–23]. We can interpret federated learning (FL) applications as a specific type of network communication problems [14]. In particular, we can apply network coding techniques to the design and analysis of federated learning (FL) methods [14].

1.1.5 Probability Theory and Statistics

Consider the data points $\mathbf{z}^{(1)}, \ldots,$ depicted in Fig. 1.2. Each data point represents some previous day that is characterized by its minimum and maximum daytime temperature as measured at a specific Finnish Meteorological Institute weather observation station. It might be useful to interpret these data points as realizations of i.i.d. random variables with common (but unknown) probability distribution $p(\mathbf{z})$. Figure 1.8 extends the scatterplot in Fig. 1.2 by adding a contour line of the underlying probability distribution $p(\mathbf{z})$ [24].

Probability theory offers principled methods for estimating the probability distribution from a set of data points (see Sect. 3.12). Given (an estimate of) the probability distribution $p(\mathbf{z})$, we can compute estimates for the label of a data point based on its features.

Having a probability distribution $p(\mathbf{z})$ for a randomly drawn data point $\mathbf{z} = (x, y)^T$, allows us to not only compute a single prediction (point estimate) $\hat{y}$ of the label y but rather an entire probability distribution $q(\hat{y})$ over all possible prediction values $\hat{y}$.

The distribution $q(\hat{y})$ represents, for each value $\hat{y}$, the probability or how likely it is that this is the true label value of the datapoint. By its very definition, this distribution $q(\hat{y})$ is precisely the conditional probability distribution $p(y|x)$ of the label value y, given the feature value x of a randomly drawn datapoint $\mathbf{z} = (x, y)^T \sim p(\mathbf{z})$.

Knowing (an accurate estimate of) the probability distribution $p(\mathbf{z})$ underlying the datapoints generated in an ML application not only allows us to compute predictions of labels. We can also use $p(\mathbf{z})$ to augment the available dataset by randomly drawing new datapoints from $p(\mathbf{z})$ (see Sect. 7.3). A recently popularized class of ML methods that use probabilistic models to generate synthetic data is known as **generative adversarial networks** [25].

Fig. 1.8 Each dot represents a datapoint $\mathbf{z} = (x, y)$ that is characterized by a numeric feature x and a numeric label y. We also indicate a contour-line of a probability distribution $p(\mathbf{z})$ that could be used to interpret data points as realizations of i.i.d. random variables with common probability distribution $p(\mathbf{z})$

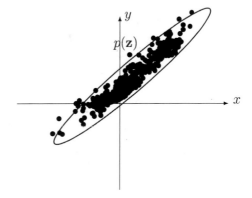

1.1.6 Artificial Intelligence

ML theory and methods are instrumental for the analysis and design of artificial intelligence [26]. An artificial intelligence system, typically referred to as an agent, interacts with its environment by executing (choosing between different) actions. These actions influence the environment as well as the state of the artificial intelligence agent. The behaviour of an artificial intelligence system is determined by how the perceptions made about the environment are used to form the next action.

From an engineering point of view, artificial intelligence aims at optimizing behaviour to maximize a long-term return. The optimization of behaviour is based solely on the perceptions made by the agent. Let us consider some application domains where AI systems can be used:

- a forest fire management system: perceptions given by satellite images and local observations using sensors or "crowd sensing" via some mobile application which allows humans to notify about relevant events; actions amount to issuing warnings and bans of open fire; return is the reduction of number of forest fires.
- a control unit for combustion engines: perceptions given by various measurements such as temperature, fuel consistency; actions amount to varying fuel feed and timing and the amount of recycled exhaust gas; return is measured in reduction of emissions.
- a severe weather warning service: perceptions given by weather radar; actions are preventive measures taken by farmers or power grid operators; return is measured by savings in damage costs (see https://www.munichre.com/)
- an automated benefit application system for the Finnish social insurance institute ("Kela"): perceptions given by information about application and applicant; actions are either to accept or to reject the application along with a justification for the decision; return is measured in reduction of processing time (applicants tend to prefer getting decisions quickly)
- a personal diet assistant: perceived environment is the food preferences of the app user and their health condition; actions amount to personalized suggestions for healthy and tasty food; return is the increase in well-being or the reduction in public spending for health-care.
- the cleaning robot Rumba (see Fig. 1.9) perceives its environment using different sensors (distance sensors, on-board camera); actions amount to choosing different moving directions ("north", "south", "east", "west"); return might be the amount of cleaned floor area within a particular time period.
- personal health assistant: perceptions given by current health condition (blood values, weight,...), lifestyle (preferred food, exercise plan); actions amount to personalized suggestions for changing lifestyle habits (less meat, more walking,...); return is measured via the level of well-being (or the reduction in public spending for health-care).
- a government-system for a country: perceived environment is constituted by current economic and demographic indicators such as unemployment rate, budget deficit, age distribution,...; actions involve the design of tax and employment

Fig. 1.9 A cleaning robot chooses actions (moving directions) to maximize a long-term reward measured by the amount of cleaned floor area per day

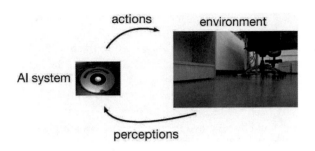

laws, public investment in infrastructure, organization of health-care system; return might be determined by the gross domestic product, the budget deficit or the gross national happiness (cf. https://en.wikipedia.org/wiki/Gross_National_Happiness).

ML methods are used on different levels by an artificial intelligence agent. On a lower level, ML methods help to extract the relevant information from raw data. ML methods are used to classify images into different categories which are then used an input for higher level functions of the artificial intelligence agent.

ML methods are also used for higher level tasks of an artificial intelligence agent. To behave optimally, an agent is required to learn a good hypothesis for how its behaviour affects its environment. We can think of optimal behaviour as a consequent choice of actions that might be predicted by ML methods.

What sets artificial intelligence applications apart from more traditional ML application is that there is an strong interaction between ML method and the data generation process. Indeed, artificial intelligence agents use the predictions of an ML method to select its next action which, in turn, influences the environment which generates new datapoints. The ML subfield of active learning studies methods that can influence the data generation [27].

Another characteristic of artificial intelligence applications is that they typically allow ML methods to evaluate the quality of a hypothesis only in hindsight. Within a basic (supervised) ML application it is possible for a ML method to try out many different hypotheses on the same data point. These different hypotheses are then scored by their discrepancies with a known correct predictions. In contrast to such passive ML applications, AI applications involve data points for which it is infeasible to determine the correct predictions.

Let us illustrate the above differences between ML and artificial intelligence applications with the help of a self-driving toy car. The toy-car is equipped with a small onboard computer, camera, sensors and actors that allow to define the steering direction. Our goal is to program the onboard computer such that it implements an artificial intelligence agent that optimally steers the toy-car. This artificial intelligence application involves data points that represent the different (temporal) states of the toy car during its ride. We use a ML method to predict the optimal steering direction for the current state. The prediction for the optimal steering angle is obtained by a hypothesis map that reads a snapshot from an on-board camera. Since these predictions are used to actually steer the car, they influence the future data points (states) that will be obtained.

Note that we typically do not know the actual optimal steering direction for each possible state of the car. It is infeasible to let the toy car roam around along any possible path and then manually label each on-board camera snapshot with the optimal steering direction (see Fig. 1.12). The usefulness of a prediction can be measured only in an indirect fashion by using some form of reward signal. Such a reward signal could be obtained from a distance sensor that allows to determine if the toy car reduced the distance to a target location.

1.2 Flavours of Machine Learning

ML methods read in data points which are generated within some application domain. An individual data point is characterized by various properties. We find it convenient to divide the properties of data points into two groups: features and labels (see Sect. 2.1). Features are properties that we measure or compute easily in an automated fashion. Labels are properties that cannot be measured easily and often represent some higher level fact (or quantity of interest) whose discovery often requires human experts.

Roughly speaking, ML aims at learning to predict (approximating or guessing) the label of a data point based solely on the features of this datapoint. Formally, the prediction is obtained as the function value of a hypothesis map whose input argument are the features of a datapoint. Since any ML method must be implemented with finite computational resources, it can only consider a subset of all possible hypothesis maps. This subset is referred to as the hypothesis space or model underlying a ML method. Based on how ML methods assess the quality of different hypothesis maps we distinguish three main flavours of ML: supervised, unsupervised and reinforcement learning.

1.2.1 Supervised Learning

The main focus of this book is on supervised ML methods. These methods use a training set that consists of labeled data points (for which we know the correct label values). We refer to a data point as labeled if its label value is known. Labeled data points might be obtained from human experts that annotate ("label") data points with their label values. There are marketplaces for renting human labelling workforce [28]. Supervised ML searches for a hypothesis that can imitate the human annotator and allows to predict the label solely from the features of a data point.

Figure 1.10 illustrates the basic principle of supervised ML methods. These methods learn a hypothesis with minimum discrepancy between its predictions and the true labels of the datapoint in the training set. Loosely speaking, supervised ML fits

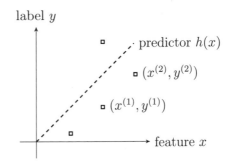

Fig. 1.10 Supervised ML methods fit a (typically highly non-linear) curve to a (typically large) set of data points

a curve (the graph of the predictor map) to labeled data points in a training set. For the actual implementing of this curve fitting we need a loss function that quantifies the fitting error. Supervised ML method differ in their choice for a loss function to measure the discrepancy between predicted label and true label of a data point.

While the principle behind supervised ML sounds trivial, the challenge of modern ML applications is the sheer amount of data points and their complexity. ML methods must process billions of data points with each single datapoint characterized by a potentially vast number of features. Consider datapoints representing social network users, whose features include all media that has been posted (videos, images, text). Besides the size and complexity of datasets, another challenge for modern ML methods is that they must be able to fit highly non-linear predictor maps. Deep learning methods address this challenge by using a computationally convenient representation of non-linear maps via artificial neural networks [10].

1.2.2 Unsupervised Learning

Some ML methods do not require knowing the label value of any data point and are therefore referred to as unsupervised ML methods. Unsupervised methods must rely solely on the intrinsic structure of data points to learn a good hypothesis. Thus, unsupervised methods do not need a teacher or domain expert who provides labels for data points (used to form a training set). Chapters 8 and 9 discuss two large families of unsupervised methods, referred to as clustering and feature learning methods.

Clustering methods group data points into few subsets such that data points within the same subset or cluster are more similar with each other than with data points outside the cluster (see Figure 1.11). Feature learning methods determine numeric features such that data points can be processed efficiently using these features. Two important applications of feature learning are dimensionality reduction and data visualization.

Fig. 1.11 Clustering
methods learn to predict the
cluster (or group)
memberships of data points
based solely on their
features. Chapter 8 discusses
clustering methods that are
unsupervised in the sense of
not requiring the knowledge
of the true cluster
membership of any data
point

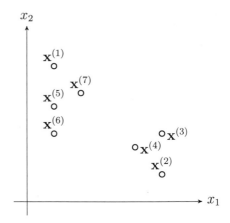

1.2.3   Reinforcement Learning

In general, ML methods use a loss function to evaluate and compare different hypotheses. The loss function assigns a (typically non-negative) loss value to a pair of a data point and a hypothesis. ML methods search for a hypothesis, out of (typically large) hypothesis space, that incurs minimum loss for any data point. Reinforcement learning (RL) studies applications where the predictions obtained by a hypothesis influences the generation of future data points. RL applications involve data points that represent the states of a (programmable) system (an artificial intelligence agent) at different time instants. The label of such a data point has the meaning of an optimal action that the agent should take in a given state. Similar to unsupervised ML, RL methods must learn a hypothesis without having access to labeled data points.

What sets RL methods apart from supervised and unsupervised methods is that it not possible for them to evaluate the loss function for different choices of a hypothesis. Consider a RL method that has to predict the optimal steering angle of a car. Naturally, we can only evaluate the usefulness specific combination of predicted label (steering angle) and the current state of the car. It is impossible to try out two different hypotheses at the same time as the car cannot follow two different steering angles (obtained by the two hypotheses) at the same time.

Mathematically speaking, RL methods can evaluate the loss function only point-wise, i.e., for the current hypothesis that has been used to obtain the most recent prediction. These point-wise evaluations of the loss function are typically implemented by using some reward signal [29]. Such a reward signal might be obtained from a sensing device and allows to quantify the usefulness of the current hypothesis.

One important application domain for RL methods is autonomous driving (see Fig. 1.12). Consider data points that represent individual time instants $t = 0, 1, \ldots$ during a car ride. The features of the tth data point are the pixel intensities of an on-board camera snapshot taken at time t. The label of this data point is the optimal steering direction at time t to maximize the distance between the car and any obstacle.

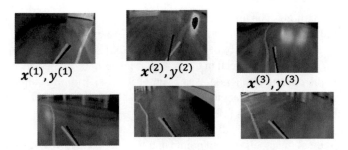

$x^{(1)}, y^{(1)}$ $x^{(2)}, y^{(2)}$

$x^{(3)}, y^{(3)}$

Fig. 1.12 Autonomous driving requires to predict the optimal steering direction (label) based on an on-board camera snapshot (features) in each time instant. RL methods sequentially adjust a hypothesis for predicting the steering direction from the snapshot. The quality of the current hypothesis is evaluated by the measurement of a distance sensor (to avoid collisions with obstacles)

We could use a ML method to learn hypothesis for predicting the optimal steering direction solely from the pixel intensities in the on-board camera snapshot. The loss incurred by a particular hypothesis is determined from the measurement of a distance sensor after the car moved along the predicted direction. We can evaluate the loss only for the hypothesis that has actually been used to predict the optimal steering direction. It is impossible to evaluate the loss for other predictions of the optimal steering direction since the car already moved on.

1.3 Organization of this Book

Chapter 2 introduces the notions of data, a model and a loss function as the three main components of ML. We will also highlight some of the computational and statistical aspects that might guide the design choices arising for these three components. A guiding theme of this book is the depiction of ML methods as combinations of specific design choices for data representation, the model and the loss function. Put differently, we aim at mapping out the vast landscape of ML methods in an abstract three-dimensional space spanned by the three dimensions: data, model and loss.

Chapter 3 details how several well-known ML methods are obtained by specific design choices for data (representation), model and loss function. Examples range from basic linear regression (see Sect. 3.1) via support vector machine (see Sect. 3.7) to deep reinforcement learning (see Sect. 3.14).

Chapter 4 discusses a principle approach to combine the three components within a practical ML method. In particular, Chap. 4 explains how a simple probabilistic model for data lends naturally to the principle of empirical risk minimization. This principle translates the problem of learning into an optimization problem. ML methods based on the empirical risk minimization are therefore a special class of optimization methods. The empirical risk minimization principle can be interpreted as a precise mathematical formulation of the "learning by trial and error" paradigm.

Chapter 5 discusses a family of iterative methods for solving the empirical risk minimization problem introduced in Chap. 4. These methods use gradients to locally approximate the objective function used in empirical risk minimization. Some variants of these gradient-based methods are currently the de-facto standard method for training deep neural networks [10].

The empirical risk minimization principle of Chap. 4 delivers a hypothesis that optimally predicts the labels of data points in a training set. However, we would like to learn a hypothesis the delivers accurate predictions also for data points that do not belong to the training set. Chapter 6 discusses some basic validation techniques that allow to probe a hypothesis outside the training set that has been used to learn (optimize) this hypothesis via empirical risk minimization. Validation techniques are instrumental for model selection, i.e., to choose the best model from a given set of candidate models. Chapter 7 presents regularization techniques that aim at replacing the training error of a candidate hypothesis with an estimate (or approximation) of its average loss incurred for data points outside the training set.

The focus of Chaps. 3–7 is on supervised ML methods that require a training set of labeled data points. Chapters 8 and 9 are devoted to unsupervised ML methods which do not require any labeled data. Chapter 8 presents some basic methods for **clustering** data. These methods group or partition data points into coherent groups which are referred to as clusters. Chapter 9 discusses **feature learning** methods that automatically determine the most relevant characteristics (or features) of a data point. This chapter also highlights the importance of using only the most relevant features of a data point, and to avoid irrelevant features, to reduce computational complexity and improve the accuracy of ML methods (such as those discussed in Chap. 3).

The successful deployment of the ML methods, such as those discussed in Chap. 3, often depends on their explainability or transparency. Chapter 10 discusses two different approaches to obtain explainable machine learning. These techniques take into account the individual user background knowledge. By analyzing a user feedback signal, which are provided for the data points in a training set, these techniques either compute personalized explanations for a given ML method or choose models that are intrinsically explainable to the user.

Prerequisites. We assume familiarity with basic notions and concepts of linear algebra, real analysis, and probability theory. For a review of those concepts, we recommend [10, Chapter 2–4] and the references therein. A main goal of this book is to develop the basic ideas and principles behind ML methods using a minimum of probability theory. However, some rudimentary knowledge about probability distributions of arbitrary random variables, probability density functions of random variables defined on Euclidean space $\mathbb{R}^n$ and probability mass functions for discrete random variables is helpful [24].

References

1. T. Mitchell, The need for biases in learning generalizations. Technical Report CBM-TR 5-110 (Rutgers University, New Brunswick, 1980)
2. W. Rudin, *Principles of Mathematical Analysis*, 3rd edn. (McGraw-Hill, New York, 1976)
3. G. Strang, *Introduction to Linear Algebra*, 5th edn. (Wellesley-Cambridge Press, Wellesley, MA, 2016)
4. S. Sra, S. Nowozin, S.J. Wright (eds.), *Optimization for Machine Learning* (MIT Press, Cambridge, 2012)
5. L. Pitt, L.G. Valiant, Computational limitations on learning from examples. J. ACM **35**(4), 965–984 (1988)
6. L.G. Valiant, A theory of the learnable, in *Proceedings of the Sixteenth Annual ACM Symposium on Theory of Computing, STOC '84* (Association for Computing Machinery, New York, 1984), pp. 436–445
7. C. Millard (ed.), *Cloud Computing Law* 2nd edn. (Oxford University Press, Oxford, 2021)
8. G.H. Golub, C.F. Van Loan, *Matrix Computations*, 3rd edn. (Johns Hopkins University Press, Baltimore, MD, 1996)
9. G. Strang, *Computational Science and Engineering* (Wellesley-Cambridge Press, Wellesley, MA, 2007)
10. I. Goodfellow, Y. Bengio, A. Courville, *Deep Learning* (MIT Press, Cambridge, MA, 2016)
11. N. Tishby, N. Zaslavsky, Deep learning and the information bottleneck principle, in *2015 IEEE Information Theory Workshop (ITW)* (IEEE, New York, 2015), pp. 1–5
12. C.E. Shannon, Communication in the presence of noise (1948)
13. T.M. Cover, J.A. Thomas, *Elements of Information Theory*, 2nd edn. (Wiley, Hoboken, NJ, 2006)
14. A.E. Gamal, Y.-H. Kim, *Network Information Theory* (Cambridge University Press, New York, 2012)
15. W. Wang, M.J. Wainwright, K. Ramchandran, Information-theoretic bounds on model selection for Gaussian Markov random fields, in *Proc. IEEE ISIT-2010* (IEEE, New York, 2010), pp. 1373–1377
16. M.J. Wainwright, Information-theoretic limits on sparsity recovery in the high-dimensional and noisy setting. IEEE Trans. Inf. Theory **55**(12), 5728–5741 (2009). (Dec.)
17. N.P. Santhanam, M.J. Wainwright, Information-theoretic limits of selecting binary graphical models in high dimensions. IEEE Trans. Inf. Theory **58**(7), 4117–4134 (2012). (Jul.)
18. N. Tran, O. Abramenko, A. Jung, On the sample complexity of graphical model selection from non-stationary samples. IEEE Transactions on Signal Processing **68**, 17–32 (2020)
19. A. Jung, Y. Eldar, N. Görtz, On the minimax risk of dictionary learning. IEEE Trans. Inf. Theory **62**(3), 1501–1515 (2016). (Mar.)
20. B. McMahan, E. Moore, D. Ramage, S. Hampson, B.A. y Arcas, Communication-efficient learning of deep networks from decentralized data, in *Proceedings of the 20th International Conference on Artificial Intelligence and Statistics*, ed. by A. Singh, J. Zhu, vol. 54 of *Proceedings of Machine Learning Research* (PMLR, 2017), pp. 1273–1282
21. V. Smith, C.-K. Chiang, M. Sanjabi, A. Talwalkar, Federated multi-task learning, in *Advances in Neural Information Processing Systems*, vol. 30, (MIT Press, Cambridge, MA, 2017)
22. F. Sattler, K. Müller, W. Samek, Clustered federated learning: Model-agnostic distributed multitask optimization under privacy constraints, in *IEEE Transactions on Neural Networks and Learning Systems* (IEEE, New York, 2020)
23. Y. SarcheshmehPour, M. Leinonen, A. Jung, Federated learning from big data over networks, in *Proceedings of the IEEE International Conferences on Acoustics, Speech and Signal Processing (ICASSP)*. Preprint at https://arxiv.org/pdf/2010.14159.pdf
24. D. Bertsekas, J. Tsitsiklis, *Introduction to Probability*, 2nd edn. (Athena Scientific, Belmont, 2008)
25. I.J. Goodfellow, J. Pouget-Abadie, M. Mirza, B. Xu, D. Warde-Farley, S. Ozair, A. Courville, Y. Bengio, Generative adversarial nets, in *Proc. Neural Inf. Proc. Syst. (NIPS)* (2014)

26. S.J. Russel, P. Norvig, *Artificial Intelligence: A Modern Approach*, 3rd edn. (Prentice Hall, New York, 2010)
27. D. Cohn, Z. Ghahramani, M. Jordan, Active learning with statistical models. J. Artif. Int. Res. **4**(1), 129–145 (1996). (March)
28. A. Sorokin, D. Forsyth, Utility data annotation with amazon mechanical turk, in *2008 IEEE Computer Society Conference on Computer Vision and Pattern Recognition Workshops* (IEEE, New York, 2008), pp. 1–8
29. R. Sutton, A. Barto, *Reinforcement Learning: An Introduction*, 2nd edn. (MIT press, Cambridge, MA, 2018)

Chapter 2
Components of ML

This book portrays ML as combinations of three components (see Fig. 2.1):

- **data** as collections of individual data points that are characterized by features (see Sect. 2.1.1) and labels (see Sect. 2.1.2)
- a model or hypothesis space that consists of computationally feasible hypothesis maps from feature space to label space (see Sect. 2.2)
- a loss function (see Sect. 2.3) to measure the quality of a hypothesis map.

A ML problem involves specific design choices for data points, its features and labels, the hypothesis space and the loss function to measure the quality of a particular hypothesis. Similar to ML problems (or applications), we can also characterize ML methods as combinations of the three above components.

We detail in Chap. 3 how some of the most popular ML methods, including linear regression (see Sect. 3.1) as well as deep learning methods (see Sect. 3.11), are obtained by specific design choices for the three components. This chapter discusses in some depth the role of and the individual components of ML and their combination in ML methods.

2.1 The Data

Data as Collections of Data points. Maybe the most important component of any ML problem (and method) is data. We consider data as collections of individual data points which are atomic units of "information containers". Data points can represent text documents, signal samples of time series generated by sensors, entire time series generated by collections of sensors, frames within a single video, random variables, videos within a movie database, cows within a herd, individual trees within a forest,

Fig. 2.1 ML methods fit a
model to data via minimizing
a loss function

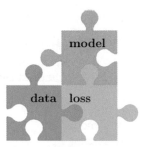

Fig. 2.2 Snapshot taken at
the beginning of a mountain
hike

individual forests within a collection of forests. Mountain hikers might be interested
in datapoints that represent different hiking tours (see Fig. 2.2).

We use the concept of datapoints in a very abstract and therefore highly flexible
manner. Datapoints can represent very different types of objects that arise in fundamentally different application domains. For an image processing application it might
be useful to define datapoints as images. When developing a recommendation system
we might define datapoints to represent customers. In the development of new drugs
we might use data points to represent different diseases. The view in this book is
that the meaning of definition of datapoints should be considered as a design choice.
We might refer to the task of finding a useful definition of data points as "data point
engineering".

One practical requirement for a useful definition of data points is that we should
have access to many of them. Many ML methods construct estimates for a quantity
of interest (such as a prediction or forecast) by averaging over a set of reference
(or training) datapoints. These estimates become more accurate for an increasing
number of datapoints used for computing the average. A key parameter of a dataset
is the number m of individual datapoints it contains. The number of datapoints within
a dataset is also referred to as the sample size. Statistically, the larger the sample size

m the better. However, there might be restrictions on computational resources (such as memory size) that limit the maximum sample size m that can be processed.

For most applications, it is impossible to have full access to every single microscopic property of a data point. Consider a datapoint that represents a vaccine. A full characterization of such a datapoint would require to specify its chemical composition down to level of molecules and atoms. Moreover, there are properties of a vaccine that depend on the patient who received the vaccine.

We find it useful to distinguish between two different groups of properties of a data point. The first group of properties is referred to as features and the second group of properties is referred to as labels. Depending on the application domain, we might refer to labels also as a **target** or the **output variable**. The features of a data point are sometimes also referred to as **input variables**.

The distinction between features and labels is somewhat blurry. The same property of a data point might be used as a feature in one application, while it might be used as a label in another application. As an example, consider feature learning for datapoints representing images. One approach to learn representative features of an image is to use some of the image pixels as the label or target pixels. We can then learn new features by learning a feature map that allows us to predict the target pixels.

To further illustrate the blurry distinction between features and labels, consider the problem of missing data. Assume we have a list of data points each of which is characterized by several properties that could be measured easily in principles (by sensors). These properties would be first candidates for being used as features of the datapoints. However, few of these properties are unknown (missing) for a small set of datapoints (e.g., due to broken sensors). We could then define the properties which are missing for some datapoints as labels and try to predict these labels using the remaining properties (which are known for all data points) as features. The task of determining missing values of properties that could be measured easily in principle is referred to as imputation [1].

Figure 2.3 illustrates two key parameters of a dataset. The first parameter is the sample size m, i.e., the number of individual data points that constitute the dataset. The second key parameter is the number n of features that are used to characterize an individual data point. The behaviour of ML methods often depends crucially on the ratio m/n. The performance of ML methods typically improves with increasing m/n. As a rule of thumb, we should use datasets for which $m/n \gg 1$. We will make the informal condition $m/n \gg 1$ more precise in Chap. 6.

2.1.1 Features

Similar to the definition of datapoints, also the choice of which properties to be used as their features is a design choice. In general, features are properties of a datapoint that can be computed or measured easily. However, this is a highly informal characterization since there no universal criterion for the difficulty of computing of measuring a property of datapoints. The task of choosing which properties to use

$$n$$

Year	m	d	Time	precp	snow	airtmp	mintmp	maxtmp
2020	1	2	00:00	0,4	55	2,5	-2	4,5
2020	1	3	00:00	1,6	53	0,8	-0,8	4,6
2020	1	4	00:00	0,1	51	-5,8	-11,1	-0,7
2020	1	5	00:00	1,9	52	-13,5	-19,1	-4,6
2020	1	6	00:00	0,6	52	-2,4	-11,4	-1
2020	1	7	00:00	4,1	52	0,4	-2	1,3

m

Fig. 2.3 Two main parameters of a dataset are the number (sample size) m of individual data points that constitute the dataset and the number n of features used to characterize individual datapoints. The behaviour of ML methods typically depends crucially on the ratio m/n

as features of data points might be the most challenging part in the application of ML methods. Chapter 9 discusses feature learning methods that automate (to some extend) the construction of good features.

In some application domains there is a rather natural choice for the features of a data point. For data points representing audio recording (of a given duration) we might use the signal amplitudes at regular sampling instants (e.g., using sampling frequency 44 kHz) as features. For data points representing images it seems natural to use the colour (red, green and blue) intensity levels of each pixel as a feature (see Fig. 2.4).

The feature construction for images depicted in Fig. 2.4 can be extended to other types of data points as long as they can be visualized efficiently. As a case in point, we might visualize an audio recording using an intensity plot of its spectrogram (see Fig. 2.5). We can then use the pixel RGB intensities of this intensity plot as the features for an audio recording. Using this trick we can transform any ML method for image data to an ML method for audio data. We can use the scatterplot of a data set to use ML methods for image segmentation to cluster the dataset (see Chap. 8).

Many important ML application domains generate data points that are characterized by several numeric features $x_1, \ldots, x_n$. We represent numeric features by real numbers $x_1, \ldots, x_n \in \mathbb{R}$ which might seem impractical. Indeed, digital computers cannot store a real number exactly as this would require an infinite number of bits. However, numeric linear algebra soft- and hardware allows to approximate real numbers with sufficient accuracy. The majority of ML methods discussed in this book assume that data points are characterized by real-valued features. Section 9.3 discusses methods for constructing numeric features of data points whose natural representation is non-numeric.

We assume that data points arising in a given ML application are characterized by the same number n of individual features $x_1, \ldots, x_n$. It is convenient to stack the individual features of a data point into a single feature vector

$$\mathbf{x} = \left(x_1, \ldots, x_n\right)^T.$$

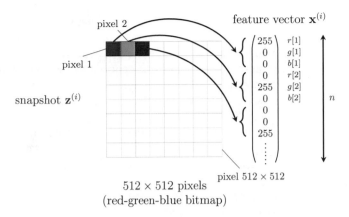

Fig. 2.4 If the snapshot $\mathbf{z}^{(i)}$ is stored as a 512×512 RGB bitmap, we could use as features $\mathbf{x}^{(i)} \in \mathbb{R}^n$ the red-, green- and blue component of each pixel in the snapshot. The length of the feature vector would then be $n = 3 \cdot 512 \cdot 512 \approx 786000$

Each data point is then characterized by its feature vector $\mathbf{x}$. Note that stacking the features of a data point into a column vector $\mathbf{x}$ is pure convention. We could also arrange the features as a row vector or even as a matrix, which might be even more natural for features obtained by the pixels of an image (see Fig. 2.4).

We refer to the set of possible feature vectors of datapoints arising in some ML application as the feature space and denote it as $\mathcal{X}$. The feature space is a design choice as it depends on what properties of a datapoint we use as its features. This design choice should take into account the statistical properties of the data as well as the available computational infrastructure. If the computational infrastructure allows for efficient numerical linear algebra, then using $\mathcal{X} = \mathbb{R}^n$ might be a good choice.

The Euclidean space $\mathbb{R}^n$ is an example of a feature space with a rich geometric and algebraic structure [2]. The algebraic structure of $\mathbb{R}^n$ is defined by vector addition and multiplication of vectors with scalars. The geometric structure of $\mathbb{R}^n$ is obtained by the Euclidean norm as a measure for the distance between two elements of $\mathbb{R}^n$. The algebraic and geometric structure of $\mathbb{R}^n$ often enables an efficient search over $\mathbb{R}^n$ to find elements with desired properties. Section 4.3 discusses examples of such search problems in the context of learning an optimal hypothesis.

Modern information-technology, including smartphones or wearables, allows us to measure a huge number of properties about datapoints in many application domains. Consider a datapoint representing the book author "Alex Jung". Alex uses a smartphone to take roughly five snapshots per day (sometimes more, e.g., during a mountain hike) resulting in more than 1000 snapshots per year. Each snapshot contains around 10^6 pixels whose greyscale levels we can use as features of the data-point. This results in more than 10^9 features (per year!). If we stack all those features into a feature vector $\mathbf{x}$, its length n would be of the order of 10^9.

As indicated above, many important ML applications involve datapoints represented by very long feature vectors. To process such high-dimensional data, modern

ML methods rely on concepts from high-dimensional statistics [3, 4]. One such concept from high-dimensional statistics is the notion of sparsity. Section 3.4 discusses methods that exploit the tendency of high-dimensional data points, which are characterized by a large number n of features, to concentrate near low-dimensional subspaces in the feature space [5].

At first sight it might seem that "the more features the better" since using more features might convey more relevant information to achieve the overall goal. However, as we discuss in Chap. 7, it can be detrimental for the performance of ML methods to use an excessive amount of (irrelevant) features. Computationally, using too many features might result in prohibitive computational resource requirements (such as processing time). Statistically, each additional feature typically introduces an additional amount of noise (due to measurement or modelling errors) which is harmful for the accuracy of the ML method.

It is difficult to give a precise and broadly applicable characterization of the maximum number of features that should be used to characterize the datapoints. As a rule of thumb, the number m of (labeled) datapoints used to train a ML method should be much larger than the number n of numeric features (see Fig. 2.3). The informal condition $m/n \gg 1$ can be ensured by either collecting a sufficiently large number m of data points or by using a sufficiently small number n of features. We next discuss implementations for each of these two complementary approaches.

The acquisition of (labeled) datapoints might be costly, requiring human expert labour. Instead of collecting more raw data, it might be more efficient to generate new artificial (synthetic) data via data augmentation techniques. Section 7.3 shows how intrinsic symmetries in the data can be used to augment the raw data with synthetic data. As an example for an intrinsic symmetry of data, consider datapoints representing an image. We assign each image the label $y = 1$ if it shows a cat and $y = -1$ otherwise. For each image with known label we can generate several augmented (additional) images with the same label. These additional images might be obtained by simple image transformation such as rotations or re-scaling (zoom-in or zoom-out) that do not change the depicted objects (the meaning of the image). Chapter 7 shows that some basic regularization techniques can be interpreted as an implicit form of data augmentation.

The informal condition $m/n \gg 1$ can also be ensured by reducing the number n of features used to characterize data points. In some applications, we might use some domain knowledge to choose the most relevant features. For other applications, it might be difficult to tell which quantities are the best choice for features. Chapter 9 discusses methods that learn, based on some given dataset, to determine a small number of relevant features of datapoints.

Beside the available computational infrastructure, also the statistical properties of datasets must be taken into account for the choices of the feature space. The linear algebraic structure of $\mathbb{R}^n$ allows us to efficiently represent and approximate datasets that are well aligned along linear subspaces. Section 9.2 discusses a basic method to optimally approximate datasets by linear subspaces of a given dimension. The geometric structure of $\mathbb{R}^n$ is also used in Chap. 8 to decompose a dataset into few groups or clusters that consist of similar data points.

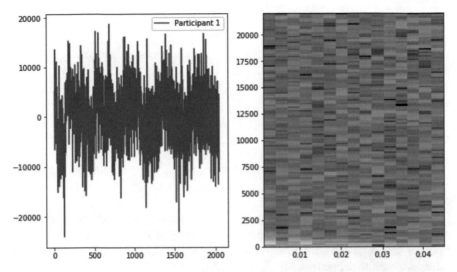

Fig. 2.5 Two visualizations of an audio recording obtained from a line plot of the signal amplitudes and by the spectrogram of the audio recording

Throughout this book we will mainly use the feature space $\mathbb{R}^n$ with dimension n being the number of features of a datapoint. This feature space has proven useful in many ML applications due to availability of efficient soft- and hardware for numerical linear algebra. Moreover, the algebraic and geometric structure of $\mathbb{R}^n$ reflect the intrinsic structure of the data generated in many important application domains. This should not be too surprising as the Euclidean space has evolved as a useful mathematical abstraction of physical phenomena.

In general there is no mathematically correct choice for which properties of a data point to be used as its features. Most application domains allow for some design freedom in the choice of features. Let us illustrate this design freedom with a personalized health-care applications. This application involves data points that represent audio recordings with the fixed duration of three seconds. These recordings are obtained via smartphone microphones and used to detect coughing [6].

Audio recordings are typically available a sequence of signal amplitudes a_t collected regularly at time instants $t = 1, \ldots, n$ with sampling frequency ≈ 44 kHz. From a signal processing perspective, it seems natural to directly use the signal amplitudes as features, $x_j = a_j$ for $j = 1, \ldots, n$. However, another choice for the features would be the pixel RGB values of some visualization of the audio recording. Figure 2.5 depicts two possible visualizations of an audio signal obtained from a line plot of the signal amplitudes (as a function of time index t) or an intensity plot of the spectrogram [7, 8].

2.1.2 Labels

Besides its features, a data point might have a different kind of properties. These properties represent a higher-level fact or quantity of interest that is associated with the data point. We refer to such properties of a data point as its label (or "output" or "target") and typically denote it by y (if it is a single number) or by $\mathbf{y}$ (if it is a vector of different label values, such as in multi-label classification). We refer to the set of all possible label values of data points arising in a ML application is the label space $\mathcal{Y}$. In general, determining the label of a data point is more difficult (to automate) compared to determining its features. Many ML methods revolve around finding efficient ways to predict (estimate or approximate) the label of a data point based solely on its features.

As already mentioned, the distinction of data point properties into labels and features is blurry. Roughly speaking, labels are properties of datapoints that might only be determined with the help of human experts. For datapoints representing humans we could define its label y as an indicator if the person has flu ($y = 1$) or not ($y = 0$). This label value can typically only be determined by a physician. However, in another application we might have enough resources to determine the flu status of any person of interest and could use it as a feature that characterizes a person.

Consider a datapoint that represents some hike, at the start of which the snapshot in Fig. 2.2 has been taken. The features of this datapoint could be the red, green and blue (RGB) intensities of each pixel in the snapshot in Fig. 2.2. We stack these RGB values into a vector $\mathbf{x} \in \mathbb{R}^n$ whose length n is three times the number of pixels in the image. The label y associated with a datapoint (which represents a hike) could be the expected hiking time to reach the mountain in the snapshot. Alternatively, we could define the label y as the water temperature of the lake visible in the snapshot.

Numeric Labels (Regression). For a given ML application, the label space $\mathcal{Y}$ contains all possible label values of data points. In general, the label space is not just a set of different elements but also equipped (algebraic or geometric) structure. To obtain efficient ML methods, we should exploit such structure. Maybe the most prominent example for such a structured label space are the real numbers $\mathcal{Y} = \mathbb{R}$. This label space is useful for ML applications involving data points with numeric labels that can be modelled by real numbers. ML methods that aim at predicting a numeric label are referred to as **regression methods**.

Categorical Labels (Classification). Many important ML applications involve data points whose label indicate the category or class to which data points belongs to. ML methods that aim at predicting such categorical labels are referred to as **classification methods**. Examples for classification problems include the diagnosis of tumours as benign or maleficent, the classification of persons into age groups or detecting the current floor conditions ("grass", "tiles" or "soil") for a mower robot.

The most simple type of a classification problems is a **binary classification** problem. Within binary classification, each datapoint belongs to exactly one out of two different classes. Thus, the label of a data point takes on values from a set that contains two different elements such as $\{0, 1\}$ or $\{-1, 1\}$ or {shows cat, shows no cat}.

We speak of a **multi-class classification** problem if data points belong to exactly one out of more than two categories (e.g., image categories "no cat shown" vs. "one cat shown" and "more than one cat shown"). If there are K different categories we might use the label values $\{1, 2, \ldots, K\}$.

There are also applications where data points can belong to several categories simultaneously. For example, an image can be cat image and a dog image at the same time if it contains a dog and a cat. Multi-label classification problems and methods use several labels $y_1, y_2, \ldots$, for different categories to which a data point can belong to. The label y_j represents the jth category and its value is $y_j = 1$ if the data point belongs to the j-th category and $y_j = 0$ if not.

Ordinal Labels. Ordinal label values are somewhat in between numeric and categorical labels. Similar to categorical labels, ordinal labels take on values from a finite set. Moreover, similar to numeric labels, ordinal labels take on values from an ordered set. For an example for such an ordered label space, consider data points representing rectangular areas of size 1 km by 1 km. The features $\mathbf{x}$ of such a data point can be obtained by stacking the RGB pixel values of a satellite image depicting that area (see Fig. 2.4). Beside the feature vector, each rectangular area is characterized by a label $y \in \{1, 2, 3\}$ where

- $y = 1$ means that the area contains no trees.
- $y = 2$ means that the area is partially covered by trees.
- $y = 3$ means that the area is entirely covered by trees.

Thus we might say that label value $y = 2$ is "larger" than label value $y = 1$ and label value $y = 3$ is "larger" than label value $y = 2$.

The distinction between regression and classification problems and methods is somewhat blurry. Consider a binary classification problem based on data points whose label y takes on values -1 or 1. We could turn this into a regression problem by using a new label y' which is defined as the confidence in the label y being equal to 1. On the other hand, given a prediction $\hat{y}'$ for the numeric label $y' \in \mathbb{R}$ we can obtain a prediction $\hat{y}$ for the binary label $y \in \{-1, 1\}$ by thresholding, $\hat{y} := 1$ if $\hat{y}' \geq 0$ whereas $\hat{y} := -1$ otherwise. A prominent example for this link between regression and classification is logistic regression which is discussed in Sect. 3.6. Despite its name, logistic regression is a binary classification method.

We refer to a data point as being *labeled* if, besides its features $\mathbf{x}$, the value of its label y is known. The acquisition of labeled data points typically involves human labour, such as handling a water thermometer at certain locations in a lake. In other applications, acquiring labels might require sending out a team of marine biologists to the Baltic sea [9], running a particle physics experiment at the European organization for nuclear research (CERN) [10], running animal testing in pharmacology [11].

Let us also point out online market places for human labelling workforce [12]. These market places, allow to upload data points, such as collections of images or audio recordings, and then offer an hourly rate to humans that label the datapoints. This labeling work might amount to marking images that show a cat.

Many applications involve datapoints whose features can be determined easily, but whose labels are known for few datapoints only. Labeled data is a scarce resource.

Some of the most successful ML methods have been devised in application domains where label information can be acquired easily [13]. ML methods for speech recognition and machine translation can make use of massive labeled datasets that are freely available [14].

In the extreme case, we do not know the label of any single datapoint. Even in the absence of any labeled data, ML methods can be useful for extracting relevant information from features only. We refer to ML methods which do not require any labeled datapoints as **unsupervised ML methods**. We discuss some of the most important unsupervised ML methods in Chaps. 8 and 9).

As discussed next, many ML methods aim at constructing (or finding) a "good" predictor $h : \mathcal{X} \to \mathcal{Y}$ which takes the features $\mathbf{x} \in \mathcal{X}$ of a datapoint as its input and outputs a predicted label (or output, or target) $\hat{y} = h(\mathbf{x}) \in \mathcal{Y}$. A good predictor should be such that $\hat{y} \approx y$, i.e., the predicted label $\hat{y}$ is close (with small error $\hat{y} - y$) to the true underlying label y.

2.1.3   Scatterplot

Consider datapoints characterized by a single numeric feature x and single numeric label y. To gain more insight into the relation between the features and label of a datapoint, it can be instructive to generate a scatterplot as shown in Fig. 1.2. A scatterplot depicts the datapoints $\mathbf{z}^{(i)} = (x^{(i)}, y^{(i)})$ in a two-dimensional plane with the axes representing the values of feature x and label y.

The visual inspection of a scatterplot might suggest potential relationships between feature x (minimum daytime temperature) and label y (maximum daytime temperature). From Fig. 1.2, it seems that there might be a relation between feature x and label y since data points with larger x tend to have larger y. This makes sense since having a larger minimum daytime temperature typically implies also a larger maximum daytime temperature.

To construct a scatterplot for data points with more than two features we can use feature learning methods (see Chap. 9). These methods transform high-dimensional datapoints, having billions of raw features, to three or two new features. These new features can then be used as the coordinates of the datapoints in a scatterplot.

2.1.4   Probabilistic Models for Data

A powerful idea in ML is to interpret each datapoints as the realization of a **random variable (RV)**. For ease of exposition let us consider datapoints that are characterized by a single feature x. The following concepts can be extended easily to datapoints characterized by a feature vector $\mathbf{x}$ and a label y.

One of the most basic examples of a probabilistic model for datapoints in ML is the i.i.d. assumption. This assumption interprets datapoints $x^{(1)}, \ldots, x^{(m)}$ as realizations of statistically independent random variables with the same probability distribution $p(x)$. It might not be immediately clear why it is a good idea to interpret data points as realizations of random variables with the common probability distribution $p(x)$. However, this interpretation allows us to use the properties of the probability distribution to characterize overall properties of entire datasets, i.e., large collections of data points.

The probability distribution $p(x)$ underlying the data points within the i.i.d. assumption is either known (based on some domain expertise) or estimated from data. It is often enough to estimate only some parameters of the distribution $p(x)$. Section 3.12 discusses a principled approach to estimate the parameters of a probability distribution from datapoints. This approach is sometimes referred to as maximum likelihood and aims at finding (parameter of) a probability distribution $p(x)$ such that the probability (density) of actually observing the available data points is maximized [15–17].

Two of the most basic and widely used parameters of a probability distribution $p(x)$ are the expected value or mean [18]

$$\mu_x = \mathbb{E}\{x\} := \int_{x'} x' p(x') dx'$$

and the variance

$$\sigma_x^2 := \mathbb{E}\{(x - \mathbb{E}\{x\})^2\}.$$

These parameters can be estimated using the sample mean (average) and sample variance,

$$\hat{\mu}_x := (1/m) \sum_{i=1}^{m} x^{(i)}, \text{ and}$$

$$\widehat{\sigma_x^2} := (1/m) \sum_{i=1}^{m} (x^{(i)} - \hat{\mu}_x)^2. \tag{2.1}$$

The sample mean and sample variance (2.1) are the maximum likelihood estimators for the mean and variance of a normal (Gaussian) distribution $p(x)$ (see [19, Chap. 2.3.4]).

Most of the ML methods discussed in this book are motivated by the i.i.d. assumption. It is important to note that this i.i.d. assumption is merely a modelling assumption. There is no means to verify if an arbitrary set of data points are "exactly" realizations of i.i.d. random variables. However, there are principled statistical methods (hypothesis tests) if a given set of data point can be well approximated as realizations of i.i.d. random variables [20]. The only way to ensure the i.i.d. assumption is to generate synthetic data using a random number generator. Such synthetic i.i.d. data points could be obtained by sampling algorithms that incrementally build a synthetic dataset by adding randomly chosen raw data points [21].

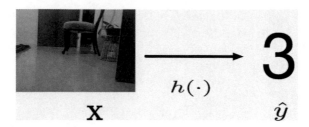

Fig. 2.6 A hypothesis (predictor) h maps features $\mathbf{x} \in \mathcal{X}$, of an on-board camera snapshot, to the prediction $\hat{y} = h(\mathbf{x}) \in \mathcal{Y}$ for the coordinate of the current location of a cleaning robot. ML methods use data to learn predictors h such that $\hat{y} \approx y$ (with true label y)

2.2   The Model

Consider some ML application that generates data points, each characterized by features $\mathbf{x} \in \mathcal{X}$ and label $y \in \mathcal{Y}$. The informal principle of most (if not every) ML method is to learn a hypothesis map $h : \mathcal{X} \to \mathcal{Y}$ such that

$$y \approx \underbrace{h(\mathbf{x})}_{\hat{y}} \text{ for any data point.} \tag{2.2}$$

The informal goal (2.2) will be made precise in several aspects throughout the rest of our book. First, we need to quantify the approximation error (2.2) incurred by a given hypothesis map h. Second, we need to make precise what we actually mean by requiring (2.2) to hold for "any" data point. We solve the first issue by the concept of a loss function in Sect. 2.3. The second issue is then solved in Chap. 4 by using a simple probabilistic model for data.

Let us assume for the time being that we have found a reasonable hypothesis h in the sense of (2.2). We can then use this hypothesis to predict the label of any data point for which we know its features. The prediction $\hat{y} = h(\mathbf{x})$ is obtained by evaluating the hypothesis for the features $\mathbf{x}$ of a data point (see Fig. 2.6. and Fig. 2.7). It seem natural to refer to a hypothesis map as a predictor map since it is used to compute predictions for the label.

For ML problems using a finite label space $\mathcal{Y}$ (e..g, $\mathcal{Y} = \{-1, 1\}$, we refer to a hypothesis also as a classifier. For a finite $\mathcal{Y}$, we can characterize a particular classifier map h using its different decision regions

$$\mathcal{R}^{(a)} := \{\mathbf{x} \in \mathbb{R}^n : h = a\} \subseteq \mathcal{X}. \tag{2.3}$$

Each label value $a \in \mathcal{Y}$ is associated with a specific decision region $\mathcal{R}^{(a)} := \{\mathbf{x} \in \mathbb{R}^n : h = a\}$. For a given label value $a \in \mathcal{Y}$, the decision region $\mathcal{R}^{(a)} := \{\mathbf{x} \in \mathbb{R}^n : h = a\}$ is constituted by all feature vectors $\mathbf{x} \in \mathcal{X}$ which are mapped to this label value, $h(\mathbf{x}) = a$.

Table 2.1 A look-up table defines a hypothesis map h. The value $h(x)$ is given by the entry in the second column of the row whose first column entry is x. We can construct a hypothesis space $\mathcal{H}$ by using a collection of different look-up tables

feature x	prediction $h(x)$
0	0
1/10	10
2/10	3
$\vdots$	$\vdots$
1	22.3

In principle, ML methods could use any possible map $h : \mathcal{X} \to \mathcal{Y}$ to predict the label $y \in \mathcal{Y}$ via computing $\hat{y} = h(\mathbf{x})$. The set of all maps from the feature space $\mathcal{X}$ to the label space is typically denoted as $\mathcal{Y}^{\mathcal{X}}$.[1] In general, the set $\mathcal{Y}^{\mathcal{X}}$ is way too large to be search over by a practical ML methods. As a point in case, consider data points characterized by a single numeric feature $x \in \mathbb{R}$ and label $y \in \mathbb{R}$. The set of all real-valued maps $h(x)$ of a real-valued argument already contains uncountably infinite many different hypothesis maps [22].

Practical ML methods can search and evaluate only a (tiny) subset of all possible hypothesis maps. This subset of computationally feasible ("affordable") hypothesis maps is referred to as the hypothesis space or model underlying a ML method. As depicted in Fig. 2.10, ML methods typically use a hypothesis space $\mathcal{H}$ that is a tiny subset of $\mathcal{Y}^{\mathcal{X}}$. Similar to the features and labels used to characterize data points, also the hypothesis space underlying a ML method is a design choice. As we will see, the choice for the hypothesis space involves a trade-off between computational complexity and statistical properties of the resulting ML methods.

The preference for a particular hypothesis space often depends on the available computational infrastructure available to a ML method. Different computational infrastructures favour different hypothesis spaces. ML methods implemented in a small embedded system, might prefer a linear hypothesis space which results in algorithms that require a small number of arithmetic operations. Deep learning methods implemented in a cloud computing environment typically use much larger hypothesis spaces obtained from deep neural networks.

ML methods can also be implemented using a spreadsheet software. Here, we might use a hypothesis space consisting of maps $h : \mathcal{X} \to \mathcal{Y}$ that are represented by look up tables (see Table 2.1). If we instead use the programming language Python to implement a ML method, we can obtain a hypothesis class by collecting all possible Python subroutines with one input (scalar feature x), one output argument (predicted label $\hat{y}$) and consisting of less than 100 lines of code.

---

[1] The notation $\mathcal{Y}^{\mathcal{X}}$ is to be understood as a symbolic shorthand and should not be understood literately as a power such as 4^5.

Broadly speaking, the design choice for the hypothesis space $\mathcal{H}$ of a ML method has to balance between two conflicting requirements.

- It has to be **sufficiently large** such that it contains at least one accurate predictor map $\hat{h} \in \mathcal{H}$. A hypothesis space $\mathcal{H}$ that is too small might fail to include a predictor map required to reproduce the (potentially highly non-linear) relation between features and label.
  Consider the task of grouping or classifying images into "cat" images and "no cat image". The classification of each image is based solely on the feature vector obtained from the pixel colour intensities. The relation between features and label ($y \in \{$cat, no cat$\}$) is highly non-linear. Any ML method that uses a hypothesis space consisting only of linear maps will most likely fail to learn a good predictor (classifier). We say that a ML method is underfitting if it uses a hypothesis space that does not contain any hypotheses maps that can accurately predict the label of any data points.
- It has to be **sufficiently small** such that its processing fits the available computational resources (memory, bandwidth, processing time). We must be able to efficiently search over the hypothesis space to find good predictors (see Sect. 2.3 and Chap. 4). This requirement implies also that the maps $h(\mathbf{x})$ contained in $\mathcal{H}$ can be evaluated (computed) efficiently [23]. Another important reason for using a hypothesis space $\mathcal{H}$ that is not too large is to avoid overfitting (see Chap. 7). If the hypothesis space $\mathcal{H}$ is too large, then just by luck we might find a hypothesis which (almost) perfectly predicts the labels of data points in a training set which is used to learn a hypothesis. However, such a hypothesis might deliver poor predictions for labels of data points outside the training set. We say that the hypothesis does not generalize well.

2.2.1   Parametrized Hypothesis spaces

A wide range of current scientific computing environments allow for efficient numerical linear algebra. This hard and software allows to efficiently process data that is provided in the form of numeric arrays such as vectors, matrices or tensors [24]. To take advantage of such computational infrastructure, many ML methods use the hypothesis space

$$\mathcal{H}^{(n)} := \{h^{(\mathbf{w})} : \mathbb{R}^n \to \mathbb{R} : h^{(\mathbf{w})}(\mathbf{x}) = \mathbf{x}^T \mathbf{w} \text{ with some weight vector } \mathbf{w} \in \mathbb{R}^n\}. \quad (2.4)$$

The hypothesis space (2.4) is constituted by linear maps (functions)

$$h^{(\mathbf{w})}(\mathbf{x}) : \mathbb{R}^n \to \mathbb{R} : \mathbf{x} \mapsto \mathbf{w}^T \mathbf{x}. \quad (2.5)$$

The function $h^{(\mathbf{w})}$ (2.5) maps, in a linear fashion, the feature vector $\mathbf{x} \in \mathbb{R}^n$ to the predicted label (or output) $h^{(\mathbf{w})}(\mathbf{x}) = \mathbf{x}^T \mathbf{w} \in \mathbb{R}$. For $n = 1$ the feature vector reduces

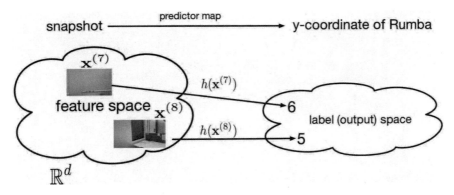

Fig. 2.7 A hypothesis $h : \mathcal{X} \to \mathcal{Y}$ takes the feature vector $\mathbf{x}^{(t)} \in \mathcal{X}$ (e.g., representing the snapshot taken by Rumba at time t) as input and outputs a predicted label $\hat{y}^{(t)} = h(\mathbf{x}^{(t)})$ (e.g., the predicted y-coordinate of Rumba at time t). A key problem studied within ML is how to automatically learn a good (accurate) predictor h such that $y^{(t)} \approx h(\mathbf{x}^{(t)})$

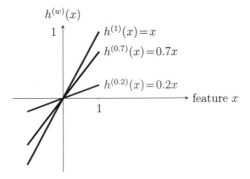

Fig. 2.8 Three particular members of the hypothesis space $\mathcal{H} = \{h^{(w)} : \mathbb{R} \to \mathbb{R}, h^{(w)}(x) = w \cdot x\}$ which consists of all linear functions of the scalar feature x. We can parametrize this hypothesis space conveniently using the weight $w \in \mathbb{R}$ as $h^{(w)}(x) = w \cdot x$

a single feature x and the hypothesis space (2.4) consists of all maps $h^{(w)}(x) = wx$ with some weight $w \in \mathbb{R}$ (see Fig. 2.8).

The elements of the hypothesis space $\mathcal{H}$ in (2.4) are parameterized by the weight vector $\mathbf{w} \in \mathbb{R}^n$. Each map $h^{(\mathbf{w})} \in \mathcal{H}$ is fully specified by the weight vector $\mathbf{w} \in \mathbb{R}^n$. This parametrization of the hypothesis space $\mathcal{H}$ allows to process and manipulate hypothesis maps by vector operations. In particular, instead of searching over the function space $\mathcal{H}$ (its elements are functions!) to find a good hypothesis, we can equivalently search over all possible weight vectors $\mathbf{w} \in \mathbb{R}^n$.

The search space $\mathbb{R}^n$ is still (uncountably) infinite but it has a rich geometric and algebraic structure that allows us to efficiently search over this space. Chapter 5 discusses methods that use the concept of gradients to implement an efficient search for good weights $\mathbf{w} \in \mathbb{R}^n$.

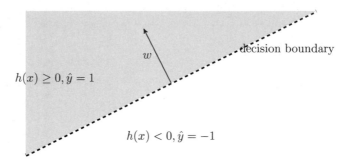

Fig. 2.9 A hypothesis $h : \mathcal{X} \to \mathcal{Y}$ for a binary classification problem, with label space $\mathcal{Y} = \{-1, 1\}$ and feature space $\mathcal{X} = \mathbb{R}^2$, can be represented conveniently via the decision boundary (dashed line) which separates all feature vectors $\mathbf{x}$ with $h(\mathbf{x}) \geq 0$ from the region of feature vectors with $h(\mathbf{x}) < 0$. If the decision boundary is a hyperplane $\{\mathbf{x} : \mathbf{w}^T \mathbf{x} = b\}$ (with normal vector $\mathbf{w} \in \mathbb{R}^n$), we refer to the map h as a linear classifier

The hypothesis space (2.4) is also appealing because of the broad availability of computing hardware such as graphic processing units. Another factor boosting the widespread use of (2.4) might be the offer for optimized software libraries for numerical linear algebra.

The hypothesis space (2.4) can also be used for classification problems, e.g., with label space $\mathcal{Y} = \{-1, 1\}$. Indeed, given a linear predictor map $h^{(\mathbf{w})}$ we can classify data points according to $\hat{y} = 1$ for $h^{(\mathbf{w})}(\mathbf{x}) \geq 0$ and $\hat{y} = -1$ otherwise. We refer to a classifier that computes the predicted label by first applying a linear map to the features as a linear classifier.

Figure 2.9 illustrates the decision regions (2.3) of a linear classifier for binary labels. The decision regions are half-spaces and, in turn, the decision boundary is a hyperplane $\{\mathbf{x} : \mathbf{w}^T \mathbf{x} = b\}$. Note that each linear classifier corresponds to a particular linear hypothesis map from the hypothesis space (2.4). However, we can use different loss functions to measure the quality of a linear classifier. Three widely-used examples for ML methods that learn a linear classifier are logistic regression (see Sect. 3.6), the support vector machine (see Sect. 3.7) and the naive Bayes classifier (see Sect. 3.8).

In some application domains, the relation between features $\mathbf{x}$ and label y of a data point is highly non-linear. As a case in point, consider data points representing images of animals. The map that relates the pixel intensities of image to the label indicating if it is a cat image is highly non-linear. For such applications, the hypothesis space (2.4) is not suitable as it only contains linear maps. The second main example for a parametrized hypothesis space studied in this book also contains non-linear maps. This parametrized hypothesis space is obtained from a parametrized signal flow diagram which is referred to as an artificial neural network. Section 3.11 will discuss the construction of non-linear parametrized hypothesis spaces using an artificial neural network.

Upgrading a Hypothesis Space via Feature Maps. Let us discuss a simple but powerful technique for enlarging ("upgrading") a given hypothesis space $\mathcal{H}$ to a larger hypothesis space $\mathcal{H}' \supseteq \mathcal{H}$ that offers a wider selection of hypothesis maps. The idea is to replace the original features $\mathbf{x}$ of a data point with new (transformed) features $\mathbf{z} = \Phi(\mathbf{x})$. The transformed features are obtained by applying a feature map $\Phi(\cdot)$ ot the original features $\mathbf{x}$. This upgraded hypothesis space $\mathcal{H}'$ consists of all concatenations of the feature map Φ and some hypothesis $h \in \mathcal{H}$,

$$\mathcal{H}' := \{h'(\cdot) : \mathbf{x} \mapsto h(\Phi(\mathbf{x})) : h \in \mathcal{H}\}. \tag{2.6}$$

The construction (2.6) used for arbitrary combinations of a feature map $\Phi(\cdot)$ and a "base" hypothesis space $\mathcal{H}$. The only requirement is that the output of the feature map can be used as input for a hypothesis $h \in \mathcal{H}$. More formally, the range of the feature map must belong to the domain of the maps in $\mathcal{H}$. Examples for ML methods that use a hypothesis space of the form (2.6) include polynomial regression (see Sect. 3.2), Gaussian basis regression (see Sect. 3.5) and the important family of kernal methods (see Sect. 3.9). The feature map in (2.6) might also be obtained from clustering or feature learning methods (see Sects. 8.4 and 9.2.1).

For the special case of the linear hypothesis space (2.4), the resulting enlarged hypothesis space (2.6) is given by all linear maps $\mathbf{w}^T \mathbf{z}$ of the transformed features $\Phi(\mathbf{x})$. Combining the hypothesis space (2.4) with a non-linear feature map results in a hypothesis space that contains non-linear maps from the original feature vector $\mathbf{x}$ to the predicted label $\hat{y}$,

$$\hat{y} = \mathbf{w}^T \mathbf{z} = \mathbf{w}^T \Phi(\mathbf{x}). \tag{2.7}$$

Non-Numeric Features. The hypothesis space (2.4) can only be used for data-points whose features are numeric vectors $\mathbf{x} = (x_1, \ldots, x_n)^T \in \mathbb{R}^n$. In some application domains, such as natural language processing, there is no obvious natural choice for numeric features. However, since ML methods based on the hypothesis space (2.4) are well developed (using numerical linear algebra), it might be useful to construct numerical features even for non-numeric data (such as text). For text data, there has been significant progress recently on methods that map a human-generated text into sequences of vectors (see [25, Chap. 12] for more details). Moreover, Sect. 9.3 will discuss an approach to generate numeric features for data points that have an intrinsic notion of similarity.

2.2.2 The Size of a Hypothesis Space

The notion of a hypothesis space being too small or being too large can be made precise in different ways. The size of a finite hypothesis space $\mathcal{H}$ can be defined as its cardinality $|\mathcal{H}|$ which is simply the number of its elements. For example, consider datapoints represented by $100 \times 10 = 1000$ black-and-white pixels and characterized by a binary label $y \in \{0, 1\}$. We can model such datapoints using the feature space

$\mathcal{X} = \{0, 1\}^{1000}$ and label space $\mathcal{Y} = \{0, 1\}$. The largest possible hypothesis space $\mathcal{H} = \mathcal{Y}^{\mathcal{X}}$ consists of all maps from $\mathcal{X}$ to $\mathcal{Y}$. The size or cardinality of this space is $|\mathcal{H}| = 2^{2^{1000}}$.

Many ML methods use a hypothesis space which contains infinitely many different predictor maps (see, e.g., (2.4)). For an infinite hypothesis space, we cannot use the number of its elements as a measure for its size. Indeed, for an infinite hypothesis space, the number of elements is not well-defined. Therefore, we measure the size of a hypothesis space $\mathcal{H}$ using its effective dimension $d_{\text{eff}}(\mathcal{H})$.

Consider a hypothesis space $\mathcal{H}$ consisting of maps $h : \mathcal{X} \rightarrow \mathcal{Y}$ that read in the features $\mathbf{x} \in \mathcal{X}$ and output an predicted label $\hat{y} = h(\mathbf{x}) \in \mathcal{Y}$. We define the effective dimension $d_{\text{eff}}(\mathcal{H})$ of $\mathcal{H}$ as the maximum number $D \in \mathbb{N}$ such that for any set $\mathcal{D} = \{(\mathbf{x}^{(1)}, y^{(1)}), \ldots, (\mathbf{x}^{(D)}, y^{(D)})\}$ of D data points with different features, we can always find a hypothesis $h \in \mathcal{H}$ that perfectly fits the labels, $y^{(i)} = h(\mathbf{x}^{(i)})$ for $i = 1, \ldots, D$.

The effective dimension of a hypothesis space is closely related to the Vapnik–Chervonenkis (VC) dimension [26]. The Vapnik–Chervonenkis (VC) dimension is maybe the most widely used concept for measuring the size of infinite hypothesis spaces [19, 26–28]. However, the precise definition of the Vapnik–Chervonenkis (VC) dimension are beyond the scope of this book. Moreover, the effective dimension captures most of the relevant properties of the Vapnik–Chervonenkis (VC) dimension for our purposes. For a precise definition of the Vapnik–Chervonenkis (VC) dimension and discussion of its applications in ML we refer to [27].

Let us illustrate our concept for the size of a hypothesis space with two examples: linear regression and polynomial regression. Linear regression uses the hypothesis space

$$\mathcal{H}^{(n)} = \{h : \mathbb{R}^n \rightarrow \mathbb{R} : h(\mathbf{x}) = \mathbf{w}^T \mathbf{x} \text{ with some vector } \mathbf{w} \in \mathbb{R}^n\}.$$

Consider a dataset $\mathcal{D} = \{(\mathbf{x}^{(1)}, y^{(1)}), \ldots, (\mathbf{x}^{(m)}, y^{(m)})\}$ consisting of m data points. We refer to this number also as the sample size of the dataset. Each data point is characterized by a feature vector $\mathbf{x}^{(i)} \in \mathbb{R}^n$ and a numeric label $y^{(i)} \in \mathbb{R}$.

Let us assume that data points are realizations of realizations of continuous i.i.d. random variables with a common probability density function. Under this assumption, the matrix

$$\mathbf{X} = (\mathbf{x}^{(1)}, \ldots, \mathbf{x}^{(m)}) \in \mathbb{R}^{n \times m},$$

which is obtained by stacking (column-wise) the feature vectors $\mathbf{x}^{(i)}$ (for $i = 1, \ldots, m$), is full rank with probability one. Basic results of linear algebra allow to show that the data points in $\mathcal{D}$ can be perfectly fit by a linear map $h \in \mathcal{H}^{(n)}$ as long as $m \leq n$. As soon as the number m of data points is not strictly larger than the number of features characterizing each data point, i.e., $m \leq n$, we can find (with probability one) a weight vector $\widehat{\mathbf{w}}$ such that $y^{(i)} = \widehat{\mathbf{w}}^T \mathbf{x}^{(i)}$ for all $i = 1, \ldots, m$. The effective dimension of the linear hypothesis space $\mathcal{H}^{(n)}$ is therefore $D = n$.

As a second example, consider the hypothesis space $\mathcal{H}_{\text{poly}}^{(n)}$ which is constituted by the set of polynomials with maximum degree n. The fundamental theorem of algebra tells us that any set of m data points with different features can be perfectly fit by a polynomial of degree n as long as $n \geq m$. Therefore, the effective dimension of

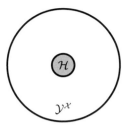

Fig. 2.10 The hypothesis space $\mathcal{H}$ is a (typically very small) subset of the (typically very large) set $\mathcal{Y}^{\mathcal{X}}$ of all possible maps from feature space $\mathcal{X}$ into the label space $\mathcal{Y}$

the hypothesis space $\mathcal{H}_{\text{poly}}^{(n)}$ is $D = n$. Section 3.2 discusses polynomial regression in more detail.

2.3 The Loss

Every ML method uses a (more of less explicit) hypothesis space $\mathcal{H}$ which consists of all **computationally feasible** predictor maps h. Which predictor map h out of all the maps in the hypothesis space $\mathcal{H}$ is the best for the ML problem at hand? To answer this questions, ML methods use the concept of a **loss function**. Formally, a loss function is a map

$$L : \mathcal{X} \times \mathcal{Y} \times \mathcal{H} \to \mathbb{R}_+ : \big((\mathbf{x}, y), h\big) \mapsto L((\mathbf{x}, y), h)$$

which assigns a pair consisting of a data point, with features $\mathbf{x}$ and label y, and a hypothesis $h \in \mathcal{H}$ the non-negative real number $L((\mathbf{x}, y), h)$.

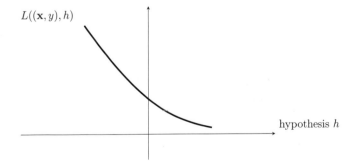

Fig. 2.11 Some loss function $L((\mathbf{x}, y), h)$ for a fixed data point, with features $\mathbf{x}$ and label y, and varying hypothesis h. ML methods try to find (learn) a hypothesis that incurs minimum loss

The loss value $L((\mathbf{x}, y), h)$ quantifies the discrepancy between the true label y and the predicted label $h(\mathbf{x})$. A small (close to zero) value $L((\mathbf{x}, y), h)$ indicates a low discrepancy between predicted label and true label of a data point. Figure 2.11 depicts a loss function for a given data point, with features $\mathbf{x}$ and label y, as a function of the hypothesis $h \in \mathcal{H}$. The basic principle of ML methods can then be formulated as: Learn (find) a hypothesis out of a given hypothesis space $\mathcal{H}$ that incurs a minimum loss $L((\mathbf{x}, y), h)$ for any data point (see Chap. 4).

Much like the choice for the hypothesis space $\mathcal{H}$ used in a ML method, also the loss function is a design choice. We will discuss some widely used examples for loss function in Sects. 2.3.1 and 2.3.2. The choice for the loss function should take into account the computational complexity of searching the hypothesis space for a hypothesis with minimum loss. Consider a ML method that uses a hypothesis space parametrized by a weight vector and a loss function that is a convex and differentiable (smooth) function of the weight vector. In this case, searching for a hypothesis with small loss can be done efficiently using the gradient-based methods discussed in Chap. 5. The minimization of a loss function that is either non-convex or non-differentiable is typically computationally much more difficult. Section 4.2 discusses the computational complexities of different types of loss functions in more detail.

Beside computational aspects, the choice of loss function should also take into account statistical aspects. Some loss functions result in ML methods that are more robust against outliers (see Sects. 3.3 and 3.7). The choice of loss function might also be guided by probabilistic models for the data generated in an ML application. Section 3.12 details how the maximum likelihood principle of statistical inference provides an explicit construction of loss functions in terms of an (assumed) probability distribution for data points.

The choice for the loss function used to evaluate the quality of a hypothesis might also be influenced by its interpretability. Section 2.3.2 discusses loss functions for hypotheses that are used to classify data points into two categories. It seems natural to measure the quality of such a hypothesis by the average number of wrongly classified data points, which is precisely the average $0/1$ loss (2.9) (see Sect. 2.3.2). Thus, the average $0/1$ loss can be interpreted as a misclassification (or error) rate. However, using the average $0/1$ loss to learn an accurate hypothesis results in computationally challenging problems. Section 2.3.2 introduces the logistic loss as a computationally attractive alternative choice for the loss function in binary classification problems.

The above aspects (computation, statistic, interpretability) result typically in conflicting goals for the choice of a loss function. A loss function that has favourable statistical properties might incur a high computational complexity of the resulting ML method. Loss functions that result in computationally efficient ML methods might not allow for an easy interpretation (what does it mean if the logistic loss of a hypothesis in a binary classification problem is 10^{-1}?). It might therefore be useful to use different loss functions for the search of a good hypothesis (see Chap. 4) and for its final evaluation. Figure 2.12 depicts an example for two such loss functions, one of them used for learning a hypothesis by minimizing the loss and the other one used for the final performance evaluation.

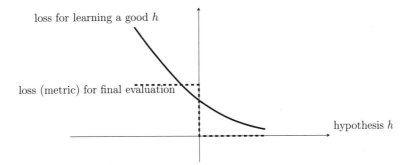

Fig. 2.12 Two different loss functions for a given data point and varying hypothesis h. One loss function (solid curve) is used to learn a good hypothesis by minimizing the loss. Another loss function (dashed curve) is used for the final performance evaluation of the learnt hypothesis. The loss function used for the final performance evaluation is referred to as a metric

For example, in a binary classification problem, we might use the logistic loss to search for (learn) an accurate hypothesis using the optimization methods in Chap. 4. The logistic loss is appealing for this purpose as it allows to efficient gradient-based methods (see Chap. 5) to search for an accurate hypothesis. After having found (learnt) an accurate hypothesis, we use the average 0/1 loss for the final performance evaluation. The 0/1 loss is appealing for this purpose as it can be interpreted as an error or misclassification rate. The loss function used for the final performance evaluation of a learnt hypothesis is sometimes referred to as metric.

2.3.1   Loss Functions for Numeric Labels

For ML problems involving data points with numeric labels $y \in \mathbb{R}$, i.e., for regression problems (see Sect. 2.1.2), a widely used (first) choice for the loss function can be the **squared error loss**

$$L((\mathbf{x}, y), h) := \big(y - \underbrace{h(\mathbf{x})}_{=\hat{y}}\big)^2. \tag{2.8}$$

The squared error loss (2.8) depends on the features $\mathbf{x}$ only via the predicted label value $\hat{y} = h(\mathbf{x})$. We can evaluate the squared error loss solely using the prediction $h(\mathbf{x})$ and the true label value y. Besides the prediction $h(\mathbf{x})$, no other properties of the features $\mathbf{x}$ are required to determine the squared error loss. We will slightly abuse notation and use the shorthand $L(y, \hat{y})$ for any loss function that depends on the features $\mathbf{x}$ only via the predicted label $\hat{y} = h(\mathbf{x})$. Figure 2.13 depicts the squared error loss as a function of the prediction error $y - \hat{y}$.

The squared error loss (2.8) has appealing computational and statistical properties. For linear predictor maps $h(\mathbf{x}) = \mathbf{w}^T \mathbf{x}$, the squared error loss is a convex and

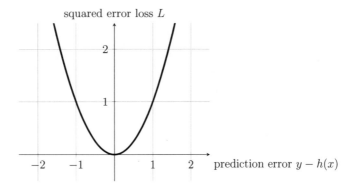

squared error loss L

prediction error $y - h(x)$

Fig. 2.13 A widely used choice for the loss function in regression problems (with data points having numeric labels) is the squared error loss (2.8). Note that, for a given hypothesis h, we can evaluate the squared error loss only if we know the features $\mathbf{x}$ and the label y of the data point

differentiable function of the weight vector $\mathbf{w}$. This allows, in turn, to efficiently search for the optimal linear predictor using efficient iterative optimization methods (see Chap. 5). The squared error loss also has a useful interpretation in terms of a probabilistic model for the features and labels. Minimizing the squared error loss is equivalent to maximum likelihood estimation within a linear Gaussian model [28, Sect. 2.6.3].

Another loss function used in regression problems is the absolute error loss $|\hat{y} - y|$. Using this loss function to guide the learning of a predictor results in methods that are robust against few outliers in the training set (see Sect. 3.3). However, this improved robustness comes at the expense of increased computational complexity of minimizing the (non-differentiable) absolute error loss compared to the (differentiable) squared error loss (2.8).

2.3.2   Loss Functions for Categorical Labels

Classification problems involve data points whose labels take on values from a discrete label space $\mathcal{Y}$. In what follows, unless stated otherwise, we focus on binary classification problems. Moreover, without loss of generality we assume that labels values are $\mathcal{Y} = \{-1, 1\}$. Classification methods aim at learning a classifier that maps the features $\mathbf{x}$ of a data point to a predicted label $\hat{y} \in \mathcal{Y}$.

We implement a classifier by thresholding the value $h(\mathbf{x}) \in \mathbb{R}$ of a hypothesis that can deliver arbitrary real numbers. We then classify a data point as $\hat{y} = 1$ if $h(\mathbf{x}) > 0$ and $\hat{y} = -1$ otherwise. Thus, the predicted label is obtained from the sign of the value $h(\mathbf{x})$. While the sign of $h(\mathbf{x})$ determines the classification result, i.e., the predicted label $\hat{y}$, we interpret the absolute value $|h(\mathbf{x})|$ as the confidence in this classification.

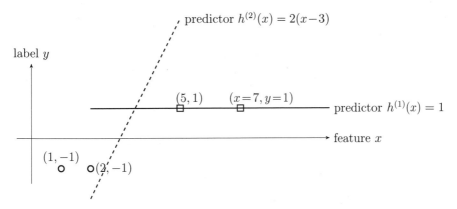

Fig. 2.14 A training set consisting of four data points with binary labels $\hat{y}^{(i)} \in \{-1, 1\}$. Minimizing the squared error loss (2.8) would prefer the (poor) classifier $h^{(1)}$ over the (reasonable) classifier $h^{(2)}$

In principle, we can measure the quality of a hypothesis when used to classify data points using the squared error loss (2.8). However, the squared error is typically a poor measure for the quality of a hypothesis $h(\mathbf{x})$ that is used to classify a data point with binary label $y \in \{-1, 1\}$. Figure 2.14 illustrates how the squared error loss of a hypothesis can be misleading in a binary classification problem.

Figure 2.14 depicts a dataset consisting of $m = 4$ data points with binary labels $y^{(i)} \in \{-1, 1\}$, for $i = 1, \ldots, m$. The figure also depicts two candidate hypotheses $h^{(1)}(x)$ and $h^{(2)}(x)$ that can be used for classifying data points. The classifications $\hat{y}$ obtained with the hypothesis $h^{(2)}(x)$ would perfectly match the labels of the four training data points since $h^{(2)}\big(x^{(i)}\big) \geq 0$ if and if only if $y^{(i)} = 1$. In contrast, the classifications $\hat{y}^{(i)}$ obtained by thresholding $h^{(1)}(x)$ are wrong for data points with $y = -1$. Thus, based on the training data, we would prefer using $h^{(2)}(x)$ over $h^{(1)}$ to classify data points. However, the squared error loss incurred by the (reasonable) classifier $h^{(2)}$ is much larger than the squared error loss incurred by the (poor) classifier $h^{(1)}$. The squared error loss is typically a bad choice for assessing the quality of a hypothesis map that is used for classifying data points into different categories.

Generally speaking, we want the loss function to punish (deliver large values for) a hypothesis that is very confident ($|h(\mathbf{x})|$ is large) in a wrong classification ($\hat{y} \neq y$). Moreover, a good loss function should not punish (deliver small values for) a hypothesis is very confident ($|h(\mathbf{x})|$ is large) in a correct classification ($\hat{y} = y$). However, by its very definition, the squared loss yields large values if the confidence $|h(\mathbf{x})|$ is large, no matter if the resulting classification is correct or wrong.

We now discuss some loss functions which have proven useful for assessing the quality of a hypothesis used to classify data points. Unless noted otherwise, the formulas for these loss functions are valid only if the label values are the real numbers -1 and 1, i.e., when the label space is $\mathcal{Y} = \{-1, 1\}$. These formulas need to modified accordingly if one prefers to use different label values for a binary classification problem. For example, instead of the label space $\mathcal{Y} = \{-1, 1\}$, we could equally well use the label space $\mathcal{Y} = \{0, 1\}$, or $\mathcal{Y} = \{\square, \triangle\}$ or $\mathcal{Y} = \{$ "Class 1", "Class 2"$\}$.

The first loss function that we discuss is direct formalization of the natural requirement for a hypothesis to result on correct classifications, i.e., $\hat{y} = y$ for any data point. This suggests to learn a hypothesis $h(\mathbf{x})$ by minimizing the 0/1 loss

$$L((\mathbf{x}, y), h) := \begin{cases} 1 & \text{if } y \neq \hat{y} \\ 0 & \text{else,} \end{cases} \text{ with } \hat{y} = 1 \text{ for } h(\mathbf{x}) \geq 0, \text{ and } \hat{y} = -1 \text{ for } h(\mathbf{x}) < 0.$$

$$(2.9)$$

Figure 2.15 illustrates the 0/1 loss (2.9) for a data point with features $\mathbf{x}$ and label $y = 1$ as a function of the hypothesis value $h(\mathbf{x})$. The 0/1 loss is equal to zero if the hypothesis yields a correct classification $\hat{y} = y$. For a wrong classification $\hat{y} \neq y$, the 0/1 loss yields the value one.

The 0/1 loss (2.9) is conceptually appealing when data points are interpreted as realizations of i.i.d. random variables with the same probability distribution $p(\mathbf{x}, y)$. Given m realizations $(\mathbf{x}^{(i)}, y^{(i)})\}_{i=1}^{m}$ of such i.i.d. random variables,

$$(1/m) \sum_{i=1}^{m} L((\mathbf{x}^{(i)}, y^{(i)}), h) \approx p(y \neq \hat{y}) \qquad (2.10)$$

with high probability for sufficiently large sample size m. A precise formulation of the approximation (2.10) can be obtained from the law of large numbers [18, Section 1]. We can apply the law of large numbers since the loss values $L((\mathbf{x}^{(i)}, y^{(i)}), h)$ are realizations of i.i.d. random variables. The average 0/1 loss on the left-hand side of (2.10) is referred to as the **accuracy** of the hypothesis h.

In view of (2.10), the 0/1 loss seems a very natural choice for assessing the quality of a classifier if our goal is to enforce correct classification ($\hat{y} = y$). This appealing statistical property of the 0/1 loss comes at the cost of high computational complexity. Indeed, for a given data point $(\mathbf{x}, y)$, the 0/1 loss (2.9) is neither convex nor differentiable when viewed as a function of the classifier h. Thus, using the 0/1 loss for binary classification problems typically involves advanced optimization methods for solving the resulting learning problem (see Sect. 3.8).

To avoid the non-convexity of the 0/1 loss we can approximate it by a convex loss function. One popular convex approximation of the 0/1 loss is the hinge loss

$$L((\mathbf{x}, y), h) := \max\{0, 1 - y \cdot h(\mathbf{x})\}. \qquad (2.11)$$

Figure 2.15 depicts the hinge loss (2.11) as a function of the hypothesis $h(\mathbf{x})$. While the hinge loss avoids the non-convexity of the 0/1 loss it still is a non-differentiable function of the classifier h. Non-differentiable loss functions are typically harder to minimize, implying a higher computational complexity of the ML method using such a loss.

Section 3.6 discusses the logistic loss which is a differentiable loss function that is useful for classification problems. The logistic loss

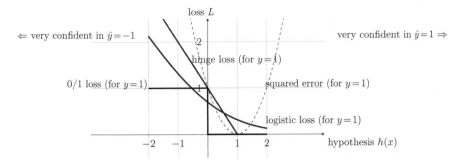

Fig. 2.15 The solid curves depict three widely-used loss functions for binary classification problems. A data point is classified as $\hat{y} = 1$ if $h(x) \geq 0$ and classified as $\hat{y} = -1$ if $h(x) < 0$. We can interpret the absolute value $|h(x)|$ as the confidence in the classification. The more confident we are in a correct classification ($\hat{y} = 1$), i.e., the more positive $h(x)$, the smaller the loss. Note that each of the three loss functions for binary classification tends monotonically to 0 for increasing $h(x)$. The dashed curve depicts the squared error loss (2.8), which increases for increasing $h(x)$

$$L((\mathbf{x}, y), h) := \log(1 + \exp(-yh(\mathbf{x}))), \tag{2.12}$$

is used within logistic regression to measure the usefulness of a linear hypothesis $h(\mathbf{x}) = \mathbf{w}^T\mathbf{x}$.

Consider a specific data point with the feature vector $\mathbf{x} \in \mathbb{R}^n$ and a binary label $y \in \{-1, 1\}$. We use a linear hypothesis $h^{(\mathbf{w})}(\mathbf{x}) = \mathbf{w}^T\mathbf{x}$, with some weight vector $\mathbf{w} \in \mathbb{R}^n$, to predict the label based on the features $\mathbf{x}$ according to $\hat{y} = 1$ if $h^{(\mathbf{w})}(\mathbf{x}) = \mathbf{w}^T\mathbf{x} > 0$ and $\hat{y} = -1$ otherwise. Then, both the hinge loss (2.11) and the logistic loss (2.12) are **convex functions** of the weight vector $\mathbf{w} \in \mathbb{R}^n$. The logistic loss (2.12) depends smoothly on $\mathbf{w}$. It is a differentiable function in the sense of allowing to define a gradient with respect to w. In contrast, the hinge loss (2.11) is nonsmooth which makes it more difficult to minimize [29, Chap. 3].

ML methods that use the convex and differentiable logistic loss function, such as logistic regression in Sect. 3.6, can apply simple gradient-based methods such as gradient descent (GD) to minimize the average loss (see Chap. 5). In contrast, we cannot use gradient-based methods to minimize the hinge loss since it is not differentiable. However, we can apply a generalization of GD which is known as subgradient descent [30] Subgradient descent is obtained from GD by generalizing the concept of a gradient to that of a subgradient.

2.3.3 Loss Functions for Ordinal Label Values

There are also loss functions particularly suited for predicting ordinal label values (see Sect. 2.1). Consider data points representing areal images of rectangular areas of size 1 km by 1 km. We characterize each data point (rectangular area) by the feature vector $\mathbf{x}$ obtained by stacking the RGB values of each image pixel (see Fig. 2.4). Beside the feature vector, each rectangular area is characterized by a label $y \in \{1, 2, 3\}$ where

- $y = 1$ means that the area contains no trees.
- $y = 2$ means that the area is partially covered by trees.
- $y = 3$ means that the area is entirely covered by trees.

Thus we might say that label value $y = 2$ is "larger" than label value $y = 1$ and label value $y = 3$ is "larger" than label value $y = 2$. It might be useful to take the ordering of label values into account when evaluating the quality of the predictions obtained by a hypothesis $h(\mathbf{x})$.

Consider a data point with feature vector $\mathbf{x}$ and label $y = 1$ as well as two different hypotheses $h^{(a)}, h^{(b)} \in \mathcal{H}$. The hypothesis $h^{(a)}$ delivers the predicted label $\hat{y}^{(a)} = h^{(a)}(\mathbf{x}) = 2$, while the other hypothesis $h^{(b)}$ delivers the predicted label $\hat{y}^{(a)} = h^{(a)}(\mathbf{x}) = 3$. Both predictions are wrong, since they are different from the true label value $y = 1$. It seems reasonable to consider the prediction $\hat{y}^{(a)}$ to be less wrong than the prediction $\hat{y}^{(b)}$ and therefore we would prefer the hypothesis $h^{(a)}$ over $h^{(b)}$. However, the 0/1 loss is the same for $h^{(a)}$ and $h^{(b)}$ and therefore does not reflect our preference for $h^{(a)}$. We need to modify (or tailor) the 0/1 loss to take into account the application-specific ordering of label values. For the above application, we might define a loss function via

$$
L((\mathbf{x}, y), h) := \begin{cases} 0 & \text{, when } y = h(\mathbf{x}) \\ 10 & \text{, when } |y - h(\mathbf{x})| = 1 \\ 100 & \text{otherwise.} \end{cases} \tag{2.13}
$$

2.3.4 Empirical Risk

The basic idea of ML methods (including those discussed in Chap. 3) is to find (or learn) a hypothesis (out of a given hypothesis space $\mathcal{H}$) that incurs minimum loss when applied to arbitrary data points. To make this informal goal precise we need to specify what we mean by "arbitrary data point". One of the most successful approaches to define the notion of "arbitrary data point" is by probabilistic models for the observed data points.

The most basic and widely-used probabilistic model interprets data points $(\mathbf{x}^{(i)}, y^{(i)})$ as realizations of i.i.d. random variables with a common probability distribution $p(\mathbf{x}, y)$. Given such a probabilistic model, it seems natural to measure the quality of a hypothesis by the expected loss or Bayes risk [15]

$$
\mathbb{E}\{L((\mathbf{x}, y), h)\} := \int_{\mathbf{x}, y} L((\mathbf{x}, y), h) dp(\mathbf{x}, y). \tag{2.14}
$$

The Bayes risk is the expected value of the loss $L((\mathbf{x}, y), h)$ incurred when applying the hypothesis h to (the realization of) a random data point with features $\mathbf{x}$ and label y. Note that the computation of the Bayes risk (2.15) requires the joint probability distribution $p(\mathbf{x}, y)$ of the (random) features and label of data points.

The Bayes risk seems to be reasonable performance measure for a hypothesis h. Indeed, the Bayes risk of a hypothesis is small only if the hypothesis incurs a small loss on average for data points drawn from the probability distribution $p(\mathbf{x}, y)$. However, it might be challenging to verify if the data points generated in a particular application domain can be accurately modelled as realizations (draws) from a probability distribution $p(\mathbf{x}, y)$. Moreover, it is also often the case that we do not know the correct probability distribution $p(\mathbf{x}, y)$.

Let us assume for the moment, that data points are generated as i.i.d. realizations of a common probability distribution $p(\mathbf{x}, y)$ which is known. It seems reasonable to learn a hypothesis h^* that incurs minimum Bayes risk,

$$\mathbb{E}\{L((\mathbf{x}, y), h^*)\} := \min_{h \in \mathcal{H}} \mathbb{E}\{L((\mathbf{x}, y), h)\}. \tag{2.15}$$

A hypothesis that solves (2.15), i.e., that achieves the minimum possible Bayes risk, is referred to as a Bayes estimator [15, Chap. 4]. The main computational challenge for learning the optimal hypothesis is the efficient (numerical) solution of the optimization problem (2.15). Efficient methods to solve the optimization problem (2.15) are studied within estimation theory [15, 31].

The focus of this book is on ML methods which do not require knowledge of the underlying probability distribution $p(\mathbf{x}, y)$. One of the most widely used principle for these ML methods is to approximate the Bayes risk by an empirical (sample) average over a finite set of labeled data $\mathcal{D} = (\mathbf{x}^{(1)}, y^{(1)}), \ldots, (\mathbf{x}^{(m)}, y^{(m)})$. In particular, we define the empirical risk of a hypothesis $h \in \mathcal{H}$ for a dataset $\mathcal{D}$ as

$$\widehat{L}(h|\mathcal{D}) = (1/m) \sum_{i=1}^{m} L((\mathbf{x}^{(i)}, y^{(i)}), h). \tag{2.16}$$

The empirical risk of the hypothesis $h \in \mathcal{H}$ is the average loss on the data points in $\mathcal{D}$. To ease notational burden, we use $\widehat{L}(h)$ as a shorthand for $\widehat{L}(h|\mathcal{D})$ if the underlying dataset $\mathcal{D}$ is clear from the context. Note that in general the empirical risk depends on both, the hypothesis h and the (features and labels of the) data points in the dataset $\mathcal{D}$.

If the data points used to compute the empirical risk (2.16) are (can be modelled as) realizations of i.i.d. random variables whose common distribution is $p(\mathbf{x}, y)$, basic results of probability theory tell us that

$$\mathbb{E}\{L((\mathbf{x}, y), h)\} \approx (1/m) \sum_{i=1}^{m} L((\mathbf{x}^{(i)}, y^{(i)}), h) \text{ for sufficiently large sample size } m.$$
$$\tag{2.17}$$

The approximation error in (2.17) can be quantified precisely by some of the most basic results of probability theory. These results are referred to as the law of large numbers.

Many (if not most) ML methods are motivated by (2.17) which suggests that a hypothesis with small empirical risk (2.16) will also result in a small expected loss. The minimum possible expected loss is achieved by the Bayes estimator of the label y, given the features $\mathbf{x}$. However, to actually compute the optimal estimator we would need to know the (joint) probability distribution $p(\mathbf{x}, y)$ of features $\mathbf{x}$ and label y.

2.3.4.1 Confusion Matrix

Consider a dataset $\mathcal{D}$ with data points characterized by feature vectors $\mathbf{x}^{(i)}$ and labels $y^{(i)} \in \{1, \ldots, k\}$. We might interpret the label value of a data point as the index of a category or class to which the data point belongs to. Multi-class classification problems aim at learning a hypothesis h such that $h(\mathbf{x}) \approx y$ for any data point.

In principle, we could measure the quality of a given hypothesis h by the average $0/1$ loss incurred on the labeled data points in (the training set) $\mathcal{D}$. However, if the dataset $\mathcal{D}$ contains mostly data points with one specific label value, the average $0/1$ loss might obscure the performance of h for data points having one of the rare label values. Indeed, even if the average $0/1$ loss is very small, the hypothesis might perform poorly for data points of a minority category.

The confusion matrix generalizes the concept of the $0/1$ loss to application domains where the relative frequency (fraction) of data points with a specific label value varies significantly (imbalanced data). Instead of considering only the average $0/1$ loss incurred by a hypothesis on a dataset $\mathcal{D}$, we use a whole family of loss functions. In particular, for each pair of label values $p, q \in \{1, \ldots, k\}$, we define the loss

$$L^{(p \to q)}\big((\mathbf{x}, y), h\big) := \begin{cases} 1 & \text{if } y = p \text{ and } h(\mathbf{x}) = q \\ 0 & \text{otherwise.} \end{cases} \tag{2.18}$$

We then compute the average loss (2.18) incurred on the dataset $\mathcal{D}$,

$$\widehat{L}^{(p \to q)}(h|\mathcal{D}) := (1/m) \sum_{i=1}^{m} L^{(p \to q)}\big((\mathbf{x}^{(i)}, y^{(i)}), h\big) \text{ for } p, q \in \{1, \ldots, k\}. \tag{2.19}$$

It is convenient to arrange the values (2.19) as a matrix which is referred to as a confusion matrix. The rows of a confusion matrix correspond to different label values p of data points. The columns of a confusion matrix correspond to different values q delivered by the hypothesis $h(\mathbf{x})$. The (p, q)-th entry of the confusion matrix is $\widehat{L}^{(p \to q)}(h|\mathcal{D})$.

2.3.4.2 Precision, Recall and F-Measure

Consider an object detection application where data points represent images. The label of data points might indicate the presence ($y = 1$)or absence ($y = -1$) of an object, it is then customary to define the [32]

$$\text{recall} := \widehat{L}^{(1 \to 1)}(h|\mathcal{D}), \text{ and the precision } := \frac{\widehat{L}^{(1 \to 1)}(h|\mathcal{D})}{\widehat{L}^{(1 \to 1)}(h|\mathcal{D}) + \widehat{L}^{(-1 \to 1)}(h|\mathcal{D})}. \tag{2.20}$$

Clearly, we would like to find a hypothesis with both, large recall and large precision. However, these two goals are typically conflicting, a hypothesis with a high recall will have small precision. Depending on the application, we might prefer having a high recall and tolerate a lower precision.

It might be convenient to combine the recall and precision of a hypothesis into a single quantity,

$$F_1 := 2 \cdot \frac{\text{precision} \cdot \text{recall}}{\text{precision} + \text{recall}} \tag{2.21}$$

The F measure (2.21) is the harmonic mean [33] of the precision and recall of a hypothesis h. It is a special case of the F_β-score

$$F_\beta := \left(1 + \beta^2\right) \cdot \frac{\text{precision} \cdot \text{recall}}{\beta^2 \text{precision} + \text{recall}}. \tag{2.22}$$

The F measure (2.21) is obtained from (2.22) for the choice $\beta = 1$. It is therefore customary to refer to (2.21) as the F_1-score of a hypothesis h.

2.3.5 Regret

In some ML applications, we might have access to the predictions obtained from some reference methods or **experts**. The quality of a hypothesis h can then be measured via the difference between the loss incurred by its predictions $h(\mathbf{x})$ and the loss incurred by the predictions of the experts [34]. This difference, which is referred to as the **regret**, measures by how much we regret to have used the prediction $h(\mathbf{x})$ instead of using (following) the prediction of the expert. The goal of regret minimization is to learn a hypothesis with a small regret compared to all considered experts.

The concept of regret minimization is useful when we do not make any probabilistic assumptions (see Sect. 2.1.4) about the data. Without a probabilistic model we cannot use the Bayes risk, which is the risk of the Bayes estimator, as a benchmark.

Regret minimization techniques can be designed and analyzed without any such probabilistic model for the data [35]. This approach replaces the Bayes risk with the regret relative to given reference predictors (experts) as the benchmark.

2.3.6 Rewards as Partial Feedback

Some applications involve data points whose labels are so difficult or costly to determine that we cannot assume to have any labeled data available. Without any labeled
data, we cannot evaluate the loss function for different choices for the hypothesis.
Indeed, the evaluation of the loss function typically amounts to measuring the distance between predicted label and true label of a data point. Instead of evaluating a
loss function, we must rely on some indirect feedback or "reward" that indicates the
usefulness of a particular prediction [35, 36].

Consider the ML problem of predicting the optimal steering directions for an
autonomous car. The prediction has to be recalculated for each new state of the
car. ML methods can sense the state via a feature vector $\mathbf{x}$ whose entries are pixel
intensities of a snapshot. The goal is to learn a hypothesis map from the feature vector
$\mathbf{x}$ to a guess $\hat{y} = h(\mathbf{x})$ for the optimal steering direction y (true label). Unless the
car circles around in small area with fixed obstacles, we have no access to labeled
datapoints or reference driving scenes for which we already know the optimum
steering direction. Instead, the car (control unit) needs to learn the hypothesis $h(\mathbf{x})$
based solely on the feedback signals obtained from various sensing devices (cameras,
distance sensors).

2.4 Putting Together the Pieces

The main theme of the book is that ML methods are obtained by different combinations of data, model and loss. We will discuss some key principles behind these
combinates in depth in the following chapters. Let us develop some intuition for
how ML methods operate by considering a very simple ML problem. This problem
involves data points that are characterized by a single numeric feature $x \in \mathbb{R}$ and a
numeric label $y \in \mathbb{R}$. We assume to have access to m labeled datapoints

$$\left(x^{(1)}, y^{(1)}\right), \ldots, \left(x^{(m)}, y^{(m)}\right) \tag{2.23}$$

for which we know the true label values $y^{(i)}$.

The assumption of knowing the exact true label values $y^{(i)}$ for any data point
is an idealization. We might often face labelling or measurement errors such that
the observed labels are noisy versions of the true label. Later on, we will discuss
techniques that allow ML methods to cope with noisy labels in Chap. 7.

Our goal is to learn a (hypothesis) map $h : \mathbb{R} \to \mathbb{R}$ such that $h(x) \approx y$ for any
data point. In other words, given any datapoint with feature x, the function value
$h(x)$ should be an accurate approximation of its label value y. We require the map
to belong to the hypothesis space $\mathcal{H}$ of linear maps,

$$h^{(w_0, w_1)}(x) = w_1 x + w_0. \tag{2.24}$$

Fig. 2.16 We can evaluate the quality of a particular predictor $h \in \mathcal{H}$ by measuring the prediction error $y - h(x)$ obtained for a labeled datapoint (x, y)

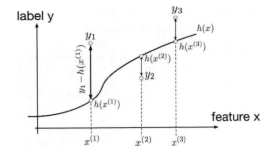

The predictor (2.24) is parameterized by the slope w_1 and the intercept (bias or offset) w_0. We indicate this by the notation $h^{(w_0, w_1)}$. A particular choice for the weights w_1, w_0 defines a linear hypothesis $h^{(w_0, w_1)}(x) = w_1 x + w_0$.

Let us use the linear hypothesis map $h^{(w_0, w_1)}(x)$ to predict the labels of training data points. In general, the predictions $\hat{y}^{(i)} = h^{(w_0, w_1)}(x^{(i)})$ will not be perfect and incur a non-zero prediction error $\hat{y}^{(i)} - y^{(i)}$ (see Fig. 2.16).

We measure the goodness of the predictor map $h^{(w_0, w_1)}$ using the average squared error loss (see (2.8))

$$f(w_0, w_1) := (1/m) \sum_{i=1}^{m} \left(y^{(i)} - h^{(w_0, w_1)}(x^{(i)}) \right)^2$$

$$\stackrel{(2.24)}{=} (1/m) \sum_{i=1}^{m} \left(y^{(i)} - (w_1 x^{(i)} + w_0) \right)^2. \tag{2.25}$$

The training error $f(w_0, w_1)$ is the average of the squared prediction errors incurred by the predictor $h^{(w_0, w_1)}(x)$ to the labeled datapoints (2.23).

It seems natural to learn a good predictor (2.24) by choosing the weights w_0, w_1 to minimize the training error

$$\min_{w_0, w_1 \in \mathbb{R}} f(w_0, w_1) \stackrel{(2.25)}{=} \min_{w_1, w_0 \in \mathbb{R}} (1/m) \sum_{i=1}^{m} \left(y^{(i)} - (w_1 x^{(i)} + w_0) \right)^2. \tag{2.26}$$

The optimal weights w_0', w_1' are characterized by the **zero-gradient condition**,[2]

$$\frac{\partial f(w_0', w_1')}{\partial w_0} = 0, \text{ and } \frac{\partial f(w_0', w_1')}{\partial w_1} = 0. \tag{2.27}$$

---

[2] A necessary and sufficient condition for $\mathbf{w}'$ to minimize a convex differentiable function $f(\mathbf{w})$ is $\nabla f(\mathbf{w}') = \mathbf{0}$ [37, Sec. 4.2.3].

Inserting (2.25) into (2.27) and by using basic rules for calculating derivatives, we obtain the following optimality conditions

$$(1/m) \sum_{i=1}^{m} \left(y^{(i)} - (w_1' x^{(i)} + w_0') \right) = 0, \text{ and } (1/m) \sum_{i=1}^{m} x^{(i)} \left(y^{(i)} - (w_1' x^{(i)} + w_0') \right) = 0. \quad (2.28)$$

Any weights w_0', w_1' that satisfy (2.28) define a predictor $h^{(w_0', w_1')} = w_1' x + w_0'$ that is optimal in the sense of incurring minimum training error,

$$f(w_0', w_1') = \min_{w_0, w_1 \in \mathbb{R}} f(w_0, w_1).$$

We find it convenient to rewrite the optimality condition (2.28) using matrices and vectors. To this end, we first rewrite the predictor (2.24) as

$$h(\mathbf{x}) = \mathbf{w}^T \mathbf{x} \text{ with } \mathbf{w} = \left(w_0, w_1 \right)^T, \mathbf{x} = \left(1, x \right)^T.$$

Let us stack the feature vectors $\mathbf{x}^{(i)} = \left(1, x^{(i)} \right)^T$ and labels $y^{(i)}$ of training datapoints (2.23) into the feature matrix and label vector,

$$\mathbf{X} = \left(\mathbf{x}^{(1)}, \ldots, \mathbf{x}^{(m)} \right)^T \in \mathbb{R}^{m \times 2}, \mathbf{y} = \left(y^{(1)}, \ldots, y^{(m)} \right)^T \in \mathbb{R}^m. \quad (2.29)$$

We can then reformulate (2.28) as

$$\mathbf{X}^T \left(\mathbf{y} - \mathbf{X}\mathbf{w}' \right) = \mathbf{0}. \quad (2.30)$$

The entries of any weight vector $\mathbf{w}' = \left(w_0', w_1' \right)$ that satisfies (2.30) are solutions to (2.28).

2.5   Exercises

Exercise 2.1 Perfect Prediction Consider data points that are characterized by a single numeric feature $x \in \mathbb{R}$ and a numeric label $y \in \mathbb{R}$. We use a ML method to learn a hypothesis map $h : \mathbb{R} \to \mathbb{R}$ based on a training set consisting of three data points

$$(x^{(1)} = 1, y^{(1)} = 3), (x^{(2)} = 4, y^{(2)} = -1), (x^{(3)} = 1, y^{(3)} = 5).$$

Is there any chance for the ML method to learn a hypothesis map that perfectly fits the training data points such that $h(x^{(i)}) = y^{(i)}$ for $i = 1, \ldots, 3$. Hint: Try to visualize the data points in a scatterplot and various hypothesis maps (see Fig. 1.3).

Exercise 2.2 Temperature Data Consider a dataset of daily air temperatures $x^{(1)}, \ldots, x^{(m)}$ measured at the observation station Utsjoki Nuorgam between 01.12.2019 and 29.02.2020. Thus, $x^{(1)}$ is the daily temperature measured on 01.12.2019, $x^{(2)}$ is the daily temperature measure don 02.12.2019, and $x^{(m)}$ is the daily temperature measured on 29.02.2020. You can download this dataset from the link https://en.ilmatieteenlaitos.fi/download-observations. ML methods often determine few parameters to characterize large collections of data points. Compute, for the above temperature measurement dataset, the following parameters

- the minimum $A := \min_{i=1,...,m} x^{(i)}$
- the maximum $B := \max_{i=1,...,m} x^{(i)}$
- the average $C := (1/m) \sum_{i=1,...,m} x^{(i)}$
- the standard deviation $D := \sqrt{(1/m) \sum_{i=1,...,m} \left(x^{(i)} - C\right)^2}$

Exercise 2.3 Deep Learning on Raspberry PI Consider the tiny desktop computer "RaspberryPI" equipped with a total of 8 Gigabytes memory [38]. On that computer, we want implement a ML algorithm that learns a hypothesis map that is represented by a deep neural network involving $n = 10^6$ numeric weights (or parameters). Each weight is quantized using 8 bits (= 1 Byte). How many different hypotheses can we store at most on a RaspberryPI computer? (You can assume that 1Gigabyte = 10^9 Bytes.)

Exercise 2.4 Ensembles. For some applications it can be a good idea to not learn a single hypothesis but to learn a whole ensemble of hypothesis maps $h^{(1)}, \ldots, h^{(B)}$. These hypotheses might even belong to different hypothesis spaces, $h^{(1)} \in \mathcal{H}^{(1)}, \ldots, h^{(B)} \in \mathcal{H}^{(B)}$. These hypothesis spaces can be arbitrary except that they are defined for the same feature space and label space. Given such an ensemble we can construct a new ("meta") hypothesis $\tilde{h}$ by combining (or aggregating) the individual predictions obtained from each hypothesis,

$$\tilde{h}(\mathbf{x}) := a\left(h^{(1)}(\mathbf{x}), \ldots, h^{(B)}(\mathbf{x})\right). \tag{2.31}$$

Here, $a(\cdot)$ denotes some given (fixed) combination or aggregation function. One example for such an aggreation function is the average $a\left(h^{(1)}(\mathbf{x}), \ldots, h^{(B)}(\mathbf{x})\right) := (1/B) \sum_{b=1}^{B} h^{(b)}(\mathbf{x})$. We obtain a new "meta" hypothesis space $\tilde{\mathcal{H}}$, that consists of all hypotheses of the form (2.31) with $h^{(1)} \in \mathcal{H}^{(1)}, \ldots, h^{(B)} \in \mathcal{H}^{(B)}$. Which conditions on the aggregation function $a(\cdot)$ and the individual hypothesis spaces $\mathcal{H}^{(1)}, \ldots, \mathcal{H}^{(B)}$ ensure that $\tilde{\mathcal{H}}$ contains each individual hypothesis space, i.e., $\mathcal{H}^{(1)}, \ldots, \mathcal{H}^{(B)} \subseteq \tilde{\mathcal{H}}$.

Exercise 2.5 How Many Features? Consider the ML problem underlying a music information retrieval smartphone app [39]. Such an app aims at identifying a song title based on a short audio recording of a song interpretation. Here, the feature vector $\mathbf{x}$ represents the sampled audio signal and the label y is a particular song title out of a huge music database. What is the length n of the feature vector $\mathbf{x} \in \mathbb{R}^n$ if its entries are the signal amplitudes of a 20-second long recording which is sampled at a rate of 44 kHz?

Exercise 2.6 Multilabel Prediction. Consider datapoints that are characterized by a feature vector $\mathbf{x} \in \mathbb{R}^{10}$ and a vector-valued label $\mathbf{y} \in \mathbb{R}^{30}$. Such vector-valued labels arise in multi-label classification problems. We want to predict the label vector using a linear predictor map

$$\mathbf{h}(\mathbf{x}) = \mathbf{W}\mathbf{x} \text{ with some matrix } \mathbf{W} \in \mathbb{R}^{30 \times 10}. \tag{2.32}$$

How many different linear predictors (2.32) are there? 10, 30, 40, or infinite?

Exercise 2.7 Average Squared Error Loss as Quadratic Form Consider the hypothesis space constituted by all linear maps $h(\mathbf{x}) = \mathbf{w}^T \mathbf{x}$ with some weight vector $\mathbf{w} \in \mathbb{R}^n$. We try to find the best linear map by minimizing the average squared error loss (the empirical risk) incurred on labeled data points (training set) $(\mathbf{x}^{(1)}, y^{(1)}), (\mathbf{x}^{(2)}, y^{(2)}), \ldots, (\mathbf{x}^{(m)}, y^{(m)})$. Is it possible to represent the resulting empirical risk as a convex quadratic function $f(\mathbf{w}) = \mathbf{w}^T \mathbf{C} \mathbf{w} + \mathbf{b} \mathbf{w} + c$? If this is possible, how are the matrix $\mathbf{C}$, vector $\mathbf{b}$ and constant c related to the feature vectors and labels of the training data?

Exercise 2.8 Find Labeled Data for Given Empirical Risk. Consider linear hypothesis space consisting of linear maps $h^{(\mathbf{w})}(\mathbf{x}) = \mathbf{w}^T \mathbf{x}$ that are parameterized by a weight vector $\mathbf{w}$. We learn an optimal weight vector by minimizing the average squared error loss $f(\mathbf{w}) = \widehat{L}(h^{(\mathbf{w})}|\mathcal{D})$ incurred by $h^{(\mathbf{w})}(\mathbf{x})$ on the training set $\mathcal{D} = (\mathbf{x}^{(1)}, y^{(1)}), \ldots, (\mathbf{x}^{(m)}, y^{(m)})$. Is it possible to reconstruct the dataset $\mathcal{D}$ just from knowing the function $f(\mathbf{w})$?. Is the resulting labeled training data unique or are there different training sets that could have resulted in the same empirical risk function? Hint: Write down the training error $f(\mathbf{w})$ in the form $f(\mathbf{w}) = \mathbf{w}^T \mathbf{Q} \mathbf{w} + c + \mathbf{b}^T \mathbf{w}$ with some matrix $\mathbf{Q}$, vector $\mathbf{b}$ and scalar c that might depend on the features and labels of the training datapoints.

Exercise 2.9 Dummy Feature Instead of Intercept Show that any hypothesis map of the form $h(x) = w_1 x + w0$ can be obtained as concatenation of a feature map $\Phi : x \mapsto \mathbf{z}$ with a map $\tilde{h}(\mathbf{z}) := \tilde{\mathbf{w}}^T \mathbf{z}$ with some weight vector $\tilde{\mathbf{w}} \in \mathbb{R}^2$.

Exercise 2.10 Approximate Non-Linear Maps Using Indicator Functions for Feature Maps. Consider an ML application generating datapoints characterized by a scalar feature $x \in \mathbb{R}$ and numeric label $y \in \mathbb{R}$. We construct a non-linear map by first transforming the feature x to a new feature vector $\mathbf{z} = (\phi_1(x), \phi_2(x), \phi_3(x), \phi_4(x))$. The components $\phi_1(x), \ldots, \phi_4(x)$ are indicator functions of intervals $[-10, -5), [-5, 0), [0, 5), [5, 10]$. In particular, $\phi_1(x) = 1$ for $x \in [-10, -5)$ and $\phi_1(x) = 0$ otherwise. We construct a hypothesis space $\mathcal{H}_1$ by all maps of the form $\mathbf{w}^T \mathbf{z}$. Note that the map is a function of the feature x since the feature vector $\mathbf{z}$ is a function of x. Which of the following predictor maps belong to $\mathcal{H}_1$?

(a)

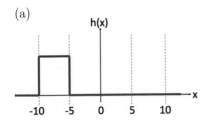

(b)

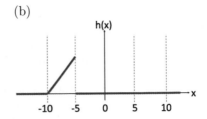

Exercise 2.11 Python Hypothesis space. Consider the source codes below for five different Python functions that read in the numeric feature x, perform some computations that result in a prediction $\hat{y}$. How large is the hypothesis space that is constituted by all maps that can be represented by one of those Python functions.

```
def func1(x):
    hat_y = 5*x+3
    return hat_y
```

```
def func2(x):
    tmp = 3*x+3
    hat_y = tmp+2*x
    return hat_y
```

```
def func3(x):
    tmp = 3*x+3
    hat_y = tmp-2*x
    return hat_y
```

```
def func4(x):
    tmp = 3*x+3
    hat_y = tmp-2*x+4
    return hat_y
```

```
def func5(x):
    tmp = 3*x+3
    hat_y = 4*tmp-2*x
    return hat_y
```

Exercise 2.12 A Lot of Features One important application domain for ML methods is healthcare. Here, data points represent human patients that are characterized by health-care records. These records might contain physiological parameters, CT scans along with various diagnoses provided by healthcare professionals. Is it a good idea to use every data field of a healthcare record as features of the data point?

Exercise 2.13 Over-Parameterization Consider datapoints characterized by feature vectors $\mathbf{x} \in \mathbb{R}^2$ and a numeric label $y \in \mathbb{R}$. We want to learn the best predictor out of the hypothesis space

$$\mathcal{H} = \{h(\mathbf{x}) = \mathbf{x}^T \mathbf{A} \mathbf{w} : \mathbf{w} \in \mathcal{S}\}.$$

Here, we used the matrix $\mathbf{A} = \begin{pmatrix} 1 & -1 \\ -1 & 1 \end{pmatrix}$ and the set

$$\mathcal{S} = \{(1, 1)^T, (2, 2)^T, (-1, 3)^T, (0, 4)^T\} \subseteq \mathbb{R}^2.$$

What is the cardinality of the hypothesis space $\mathcal{H}$, i.e., how many different predictor maps does $\mathcal{H}$ contain?

Exercise 2.14 Squared Error Loss Consider a hypothesis space $\mathcal{H}$ constituted by three predictors $h^{(1)}(\cdot), h^{(2)}(\cdot), h^{(3)}(\cdot)$. Each predictor $h^{(j)}(x)$ is a real-valued function of a real-valued argument x. Moreover, for each $j \in \{1, 2, 3\}$, $h^{(j)}(x) = 0$ for all $x^2 \leq j$ and $h^{(j)}(x) = j$ otherwise. Can you tell which of these hypothesis is optimal in the sense of having smallest average squared error loss on the three (training) datapoints $(x = 1/10, y = 3)$, $(0, 0)$ and $(1, -1)$.

Exercise 2.15 Classification Loss The Fig. 2.15 depicts different loss functions for a fixed data point with label $y = 1$ and varying hypothesis $h \in \mathcal{H}$. How would Fig. 2.15 change if we evaluate the same lloss functions for another data point $z = (x, y)$ with label $y = -1$?

Exercise 2.16 Intercept Term Linear regression methods model the relation between the label y and feature x of a datapoint as $y = h(x) + e$ with some small additive term e. The predictor map $h(x)$ is assumed to be linear $h(x) = w_1 x + w_0$. The weight w_0 is sometimes referred to as the intercept (or bias) term. Assume we know for a given linear predictor map its values $h(x)$ for $x = 1$ and $x = 3$. Can you determine the weights w_1 and w_0 based on $h(1)$ and $h(3)$?

Exercise 2.17 Picture Classification Consider a huge collection of outdoor pictures you have taken during your last adventure trip. You want to organize these pictures as three categories (or classes) *dog*, *bird* and *fish*. How could you formalize this task as a ML problem?

Exercise 2.18 Maximum Hypothesis space Consider datapoints characterized by a single real-valued feature x and a single real-valued label y. How large is the largest possible hypothesis space of predictor maps $h(x)$ that read in the feature value of a datapoint and deliver a real-valued prediction $\hat{y} = h(x)$?

Exercise 2.19 A Large but Finite Hypothesis space Consider datapoints whose features are 10×10 black-and-white (bw) pixel images. Each datapoint is also characterized by a binary label $y \in \{0, 1\}$. Consider the hypothesis space which is constituted by all maps that take a bw image as input and deliver a prediction for the label. How large is this hypothesis space?

Exercise 2.20 Size of Linear Hypothesis space Consider a training set of m datapoints with feature vectors $\mathbf{x}^{(i)} \in \mathbb{R}^n$ and numeric labels $y^{(1)}, \ldots, y^{(m)}$. The feature vectors and label values of the training set are arbitrary except that we assume the feature matrix $\mathbf{X} = (\mathbf{x}^{(1)}, \ldots)$ is full rank. What condition on m and n guarantees that we can find a linear predictor $h(\mathbf{x}) = \mathbf{w}^T \mathbf{x}$ that perfectly fits the training set, i.e., $y^{(1)} = h(\mathbf{x}^{(1)}), \ldots, y^{(m)} = h(\mathbf{x}^{(m)})$.

References

1. K. Abayomi, A. Gelman, M.A. Levy, Diagnostics for multivariate imputations. Journal of The Royal Statistical Society Series C-applied Statistics **57**, 273–291 (2008)
2. W. Rudin, *Principles of Mathematical Analysis*, 3rd edn. (McGraw-Hill, New York, 1976)
3. P. Bühlmann, S. van de Geer, *Statistics for High-Dimensional Data* (Springer, New York, 2011)
4. M. Wainwright, *High-Dimensional Statistics: A Non-Asymptotic Viewpoint* (Cambridge University Press, Cambridge, 2019)
5. R. Vidal, Subspace clustering. *IEEE Signal Processing Magazine*, March 2011
6. F. Barata, K. Kipfer, M. Weber, P. Tinschert, E. Fleisch, and T. Kowatsch, Towards device-agnostic mobile cough detection with convolutional neural networks, in *2019 IEEE International Conference on Healthcare Informatics (ICHI)*, pp. 1–11 (IEEE, New York, 2019)
7. B. Boashash (ed.), *Time Frequency Signal Analysis and Processing: A Comprehensive Reference* (Elsevier, Amsterdam, The Netherlands, 2003)
8. S.G. Mallat, *A Wavelet Tour of Signal Processing - The Sparse Way*, 3rd edn. (Academic Press, San Diego, CA, 2009)
9. S. Smoliski, K. Radtke, Spatial prediction of demersal fish diversity in the Baltic sea: comparison of machine learning and regression-based techniques. ICES Journal of Marine Science **74**(1), 102–111 (2017)
10. S. Carrazza, Machine learning challenges in theoretical HEP (2018)
11. M. Gao, H. Igata, A. Takeuchi, K. Sato, Y. Ikegaya, Machine learning-based prediction of adverse drug effects: An example of seizure-inducing compounds. Journal of Pharmacological Sciences **133**(2), 70–78 (2017)
12. K. Mortensen, T. Hughes, Comparing amazon's mechanical Turk platform to conventional data collection methods in the health and medical research literature. J. Gen. Intern Med. **33**(4), 533–538 (2018)
13. A. Halevy, P. Norvig, F. Pereira, *The unreasonable effectiveness of data* (IEEE Intelligent Systems, New York, 2009)
14. P. Koehn, Europarl: A parallel corpus for statistical machine translation, in *The 10th Machine Translation Summit*, pp. 79–86 (AAMT, 2005)
15. E.L. Lehmann, G. Casella, *Theory of Point Estimation*, 2nd edn. (Springer, New York, 1998)
16. S.M. Kay, *Fundamentals of Statistical Signal Processing: Estimation Theory* (Prentice Hall, Englewood Cliffs, NJ, 1993)
17. D. Bertsekas, J. Tsitsiklis, *Introduction to Probability*, 2nd edn. (Athena Scientific, Singapore, 2008)
18. P. Billingsley, *Probability and Measure*, 3rd edn. (Wiley, New York, 1995)
19. C.M. Bishop, *Pattern Recognition and Machine Learning* (Springer, Berlin, 2006)
20. H. Lütkepohl, *New Introduction to Multiple Time Series Analysis* (Springer, New York, 2005)
21. B. Efron, R. Tibshirani, Improvements on cross-validation: The 632+ bootstrap method. Journal of the American Statistical Association **92**(438), 548–560 (1997)
22. P. Halmos, *Naive Set Theory* (Springer, Berlin, 1974)
23. P. Austin, P. Kaski, and K. Kubjas, Tensor network complexity of multilinear maps (2018)
24. F. Pedregosa, Scikit-learn: Machine learning in python. Journal of Machine Learning Research **12**(85), 2825–2830 (2011)
25. I. Goodfellow, Y. Bengio, A. Courville, *Deep Learning* (MIT Press, Cambridge, 2016)
26. V.N. Vapnik, *The Nature of Statistical Learning Theory* (Springer, Berlin, 1999)
27. S. Shalev-Shwartz, S. Ben-David, *Understanding Machine Learning-From Theory to Algorithms* (Cambridge University Press, New York, 2014)
28. T. Hastie, R. Tibshirani, J. Friedman, *The Elements of Statistical Learning* Springer Series in Statistics. (Springer, New York, 2001)
29. Y. Nesterov, *Introductory Lectures on Convex Optimization, Vol. 87 of Applied Optimization* (Kluwer Academic Publishers, Boston, 2004)
30. S. Bubeck, Convex optimization: Algorithms and complexity. Foundations and Trends in Machine Learning **8**(3–4), 231–357 (2015)

31. M.J. Wainwright, M.I. Jordan, *Graphical Models, Exponential Families, and Variational Inference, Foundations and Trends in Machine Learning*, vol. 1 (Now Publishers, Hanover, MA, 2008)
32. R. Baeza-Yates, B. Ribeiro-Neto, *Modern Information Retrieval* (ACM Press, New York, 1999)
33. M. Abramowitz, I.A. Stegun (eds.), *Handbook of Mathematical Functions* (Dover, New York, 1965)
34. E. Hazan, *Introduction to Online Convex Optimization* (Now Publishers Inc., Hanover, MA, 2016)
35. N. Cesa-Bianchi, G. Lugosi, *Prediction, Learning, and Games* (Cambridge University Press, New York, 2006)
36. R. Sutton, A. Barto, *Reinforcement Learning: An Introduction*, 2nd edn. (MIT Press, Cambridge, MA, 2018)
37. S. Boyd, L. Vandenberghe, *Convex Optimization* (Cambridge University Press, Cambridge, UK, 2004)
38. O. Dürr, Y. Pauchard, D. Browarnik, R. Axthelm, and M. Loeser. Deep learning on a raspberry pi for real time face recognition (2015)
39. A. Wang, An industrial-strength audio search algorithm, in *International Symposium on Music Information Retrieval* (2003)

Chapter 3
The Landscape of ML

As discussed in Chap. 2, ML methods combine three main components:

- a set of data points that are characterized by features and labels
- a model or hypothesis space $\mathcal{H}$ that consists of different hypotheses $h \in \mathcal{H}$.
- a loss function to measure the quality of a particular hypothesis h.

Each of these three components involves design choices for the representation of data, their features and labels, the model and loss function. This chapter details the high-level design choices used by some of the most popular ML methods. Figure 3.1 depicts these ML methods in a two-dimensional plane whose horizontal axes represents different hypothesis spaces and the vertical axis represents different loss functions.

To obtain a practical ML method we also need to combine the above components. The basic principle of any ML method is to search the model for a hypothesis that incurs minimum loss on any data point. Chapter 4 will then discuss a principled way to turn this informal statement into actual ML algorithms that could be implemented on a computer.

3.1 Linear Regression

Consider data points characterized by feature vectors $\mathbf{x} \in \mathbb{R}^n$ and numeric label $y \in \mathbb{R}$. Linear regression aims at learning a hypothesis out of the linear hypothesis space

$$\mathcal{H}^{(n)} := \{h^{(\mathbf{w})} : \mathbb{R}^n \to \mathbb{R} : h^{(\mathbf{w})}(\mathbf{x}) = \mathbf{w}^T \mathbf{x} \text{ with some weight vector } \mathbf{w} \in \mathbb{R}^n\}. \quad (3.1)$$

Figure 1.3 depicts the graphs of some maps from $\mathcal{H}^{(2)}$ for data points with feature vectors of the form $\mathbf{x} = (1, x)^T$. The quality of a particular predictor $h^{(\mathbf{w})}$ is measured by the squared error loss (2.8). Using labeled data $\mathcal{D} = \{(\mathbf{x}^{(i)}, y^{(i)})\}_{i=1}^m$, linear

© The Author(s), under exclusive license to Springer Nature Singapore Pte Ltd. 2022
A. Jung, *Machine Learning*, Machine Learning: Foundations, Methodologies,
and Applications, https://doi.org/10.1007/978-981-16-8193-6_3

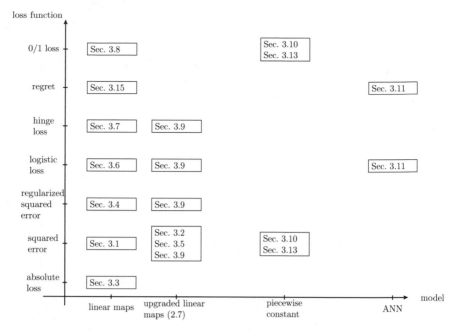

Fig. 3.1 ML methods fit a model to data by minimizing a loss function. Different ML methods use different design choices for data, model and loss

regression learns a predictor $\hat{h}$ which minimizes the average squared error loss, or **mean squared error**, (see (2.8))

$$\hat{h} = \operatorname*{argmin}_{h \in \mathcal{H}^{(n)}} \widehat{L}(h|\mathcal{D}) \stackrel{(2.16)}{=} \operatorname*{argmin}_{h \in \mathcal{H}^{(n)}} (1/m) \sum_{i=1}^{m} (y^{(i)} - h(\mathbf{x}^{(i)}))^2. \qquad (3.2)$$

Since the hypothesis space $\mathcal{H}^{(n)}$ is parameterized by the weight vector $\mathbf{w}$ (see (3.1)), we can rewrite (3.2) as an optimization problem directly over the weight vector $\mathbf{w}$:

$$\widehat{\mathbf{w}} = \operatorname*{argmin}_{\mathbf{w} \in \mathbb{R}^n} (1/m) \sum_{i=1}^{m} (y^{(i)} - h^{(\mathbf{w})}(\mathbf{x}^{(i)}))^2$$

$$\stackrel{h^{(\mathbf{w})}(\mathbf{x})=\mathbf{w}^T\mathbf{x}}{=} \operatorname*{argmin}_{\mathbf{w} \in \mathbb{R}^n} (1/m) \sum_{i=1}^{m} (y^{(i)} - \mathbf{w}^T\mathbf{x}^{(i)})^2. \qquad (3.3)$$

The optimization problems (3.2) and (3.3) are equivalent in the following sense: Any optimal weight vector $\widehat{\mathbf{w}}$ which solves (3.3), can be used to construct an optimal predictor $\hat{h}$, which solves (3.2), via $\hat{h}(\mathbf{x}) = h^{(\widehat{\mathbf{w}})}(\mathbf{x}) = (\widehat{\mathbf{w}})^T\mathbf{x}$.

Fig. 3.2 A scatterplot that depicts some data points $(x^{(1)}, y^{(1)}), \ldots,$. The ith datapoint is depicted by a dot whose coordinates are the feature $x^{(i)}$ and label $y^{(i)}$ of that datapoint

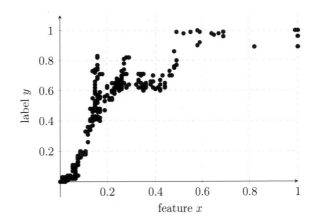

3.2 Polynomial Regression

Consider an ML problem involving datapoints which are characterized by a single numeric feature $x \in \mathbb{R}$ (the feature space is $\mathcal{X} = \mathbb{R}$) and a numeric label $y \in \mathbb{R}$ (the label space is $\mathcal{Y} = \mathbb{R}$). We observe a bunch of labeled datapoints which are depicted in Fig. 3.2.

Figure 3.2 suggests that the relation $x \mapsto y$ between feature x and label y is highly non-linear. For such non-linear relations between features and labels it is useful to consider a hypothesis space which is constituted by polynomial maps

$$\mathcal{H}_{\text{poly}}^{(n)} = \{h^{(\mathbf{w})} : \mathbb{R} \to \mathbb{R} : h^{(\mathbf{w})}(x) = \sum_{r=1}^{n} w_r x^{r-1}, \text{ with some } \mathbf{w} = (w_1, \ldots, w_n)^T \in \mathbb{R}^n\}.$$

(3.4)

We can approximate any non-linear relation $y = h(x)$ with any desired level of accuracy using a polynomial $\sum_{r=1}^{n} w_r x^{r-1}$ of sufficiently large degree n.[1]

For linear regression (see Sect. 3.1), we measure the quality of a predictor by the squared error loss (2.8). Based on labeled data points $\mathcal{D} = \{(x^{(i)}, y^{(i)})\}_{i=1}^{m}$, each having a scalar feature $x^{(i)}$ and label $y^{(i)}$, polynomial regression minimizes the average squared error loss (see (2.8)):

$$\min_{h \in \mathcal{H}_{\text{poly}}^{(n)}} (1/m) \sum_{i=1}^{m} (y^{(i)} - h^{(\mathbf{w})}(x^{(i)}))^2.$$

(3.5)

It is customary to refer to the average squared error loss also as the mean squared error.

---

[1] The precise formulation of this statement is known as the "Stone-Weierstrass Theorem" [1, Thm. 7.26].

We can interpret polynomial regression as a combination of a feature map (transformation) (see Sect. 2.1.1) and linear regression (see Sect. 3.1). Indeed, any polynomial predictor $h^{(\mathbf{w})} \in \mathcal{H}_{\text{poly}}^{(n)}$ is obtained as a concatenation of the feature map

$$\Phi(x) \mapsto (1, x, \dots, x^n)^T \in \mathbb{R}^{n+1} \tag{3.6}$$

with some linear map $\tilde{h}^{(\mathbf{w})} : \mathbb{R}^{n+1} \to \mathbb{R} : \mathbf{x} \mapsto \mathbf{w}^T \mathbf{x}$, i.e.,

$$h^{(\mathbf{w})}(x) = \tilde{h}^{(\mathbf{w})}(\Phi(x)). \tag{3.7}$$

Thus, we can implement polynomial regression by first applying the feature map Φ (see (3.6)) to the scalar features $x^{(i)}$, resulting in the transformed feature vectors

$$\mathbf{x}^{(i)} = \Phi(x^{(i)}) = \left(1, x^{(i)}, \dots, (x^{(i)})^{n-1}\right)^T \in \mathbb{R}^n, \tag{3.8}$$

and then applying linear regression (see Sect. 3.1) to these new feature vectors.

By inserting (3.7) into (3.5), we obtain a linear regression problem (3.3) with feature vectors (3.8). Thus, while a predictor $h^{(\mathbf{w})} \in \mathcal{H}_{\text{poly}}^{(n)}$ is a non-linear function $h^{(\mathbf{w})}(x)$ of the original feature x, it is a linear function $\tilde{h}^{(\mathbf{w})}(\mathbf{x}) = \mathbf{w}^T \mathbf{x}$ (see (3.7)), of the transformed features $\mathbf{x}$ (3.8).

3.3 Least Absolute Deviation Regression

Learning a linear predictor by minimizing the average squared error loss incurred on training data is not robust against the presence of outliers. This sensitivity to outliers is rooted in the properties of the squared error loss $(y - h(\mathbf{x}))^2$. Minimizing the average squared error forces the resulting predictor $\hat{y}$ to not be too far away from any datapoint. However, it might be useful to tolerate a large prediction error $y - h(\mathbf{x})$ for an unusual or exceptional data point that can be considered an outlier.

Replacing the squared loss with a different loss function can make the learning robust against outliers. One important example for such a "robustifying" loss function is the Huber loss [2]

$$L\big((\mathbf{x}, y), h\big) = \begin{cases} (1/2)(y - h(\mathbf{x}))^2 & \text{for } |y - h(\mathbf{x})| \le \varepsilon \\ \varepsilon(|y - h(\mathbf{x})| - \varepsilon/2) & \text{else.} \end{cases} \tag{3.9}$$

Figure 3.3 depicts the Huber loss as a function of the prediction error $y - h(\mathbf{x})$.

The Huber loss definition (3.9) contains a tuning parameter ϵ. The value of this tuning parameter defines when a data point is considered as an outlier. Figure 3.4 illustrates the role of this parameter as the width of a band around a hypothesis map. The prediction error of this hypothesis map for data points within this band are

Fig. 3.3 The Huber loss (3.9) resembles the squared error loss (2.8) for small prediction error and the absolute difference loss for larger prediction errors

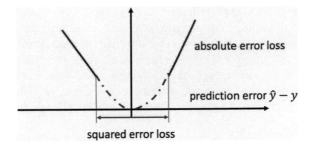

Fig. 3.4 The Huber loss measures prediction errors via squared error loss for regular data points inside the band of width ε around the hypothesis map $h(\mathbf{x})$ and via the absolute difference loss for an outlier outside the band

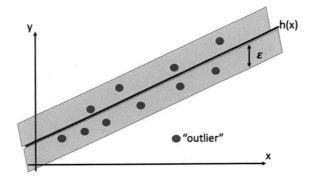

measured used squared error loss (2.8). For data points outside this band (outliers) we use instead the absolute value of the prediction error as the resulting loss.

The Huber loss is robust to outliers since the corresponding (large) prediction errors $y - \hat{y}$ are not squared. Outliers have a smaller effect on the average Huber loss (over the entire dataset) compared to the average squared error loss. The improved robustness against outliers of the Huber loss comes at the expense of increased computational complexity. The squared error loss can be minimized using efficient gradient based methods (see Chap. 5). In contrast, for $\varepsilon = 0$, the Huber loss is non-differentiable and requires more advanced optimization methods.

The Huber loss (3.9) contains two important special cases. The first special case occurs when ε is chosen to be very large, such that the condition $|y - \hat{y}| \le \varepsilon$ is satisfied for most datapoints. In this case, the Huber loss resembles the squared error loss (2.8) (up to a scaling factor $1/2$). The second special case is obtained for $\varepsilon = 0$. In this case, the Huber loss reduces to the scaled absolute difference loss $|y - \hat{y}|$.

3.4 The Lasso

We will see in Chap. 6 that linear regression (see Sect. 3.1) typically requires a training set larger than the number of features used to characterized a data point. However, many important application domains generate data points with a number

n of features much higher than the number m of available labeled data points in the training set. In this high-dimensional regime, where $m \ll n$, basic linear regression will not be able to learn useful weights $\mathbf{w}$ for a linear hypothesis.

Section 6.4 shows that for $m \ll n$, linear regression will typically learn a hypothesis that perfectly predicts labels of data points in the training set but delivers poor predictions for data points outside the training set. This phenomenon is referred to as overfitting and poses a main challenge for ML applications in the high-dimensional regime.

Chapter 7 discusses basic regularization techniques that allow to prevent ML methods from overfitting. We can regularize linear regression by augmenting the squared error loss (2.8) of a hypothesis $h^{(\mathbf{w})}(\mathbf{x}) = \mathbf{w}^T \mathbf{x}$ with an additional penalty term. This penalty term depends solely on the weights $\mathbf{w}$ and serves as an estimate for the increase of the average loss on data points outside the training set. Different ML methods are obtained from different choices for this penalty term. The least absolute shrinkage and selection operator (Lasso) is obtained from linear regression by replacing the squared error loss with the regularized loss

$$L((\mathbf{x}, y), h^{(\mathbf{w})}) = (y - \mathbf{w}^T \mathbf{x})^2 + \lambda \|\mathbf{w}\|_1. \tag{3.10}$$

Here, the penalty term is given by the scaled norm $\lambda \|\mathbf{w}\|_1$. The value of λ can be chosen based on some probabilistic model that interprets a data point as the realization of a random variable. The label of this random datapoint is related to its features via

$$y = \overline{\mathbf{w}}^T \mathbf{x} + \varepsilon.$$

Here, $\overline{\mathbf{w}}$ denotes some true underlying weight vector and ε is a realization of an a random variable that is independent of the features $\mathbf{x}$. We need the "noise" term ε since the labels of datapoints collected in some ML application are typically not exactly obtained by a linear combination $\overline{\mathbf{w}}^T \mathbf{x}$ of its features.

The tuning of λ in (3.10) can be guided by the statistical properties (such as the variance) of the noise ε, the number of non-zero entries in $\overline{\mathbf{w}}$ and a lower bound on the non-zero values [3, 4]. Another option for choosing the value λ is to try out different candidate values and pick the one resulting in smallest validation error (see Sect. 6.2).

3.5   Gaussian Basis Regression

Section 3.2 showed how to extend linear regression by first transforming the feature x using a vector-valued feature map $\Phi : \mathbb{R} \to \mathbb{R}^n$. The output of this feature map are the transformed features $\Phi(x)$ which are fed, in turn, to a linear map $h(\Phi(x)) = \mathbf{w}^T \Phi(x)$. Polynomial regression in Sect. 3.2 has been obtained for the specific feature map (3.6) whose entries are the powers x^l of the scalar original feature x. However, it is possible to use other functions, different from polynomials, to construct the feature

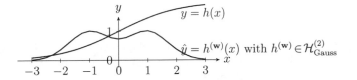

Fig. 3.5 The true relation $x \mapsto y$ (blue) between feature x and label y of data points is highly non-linear. Therefore it seems reasonable to predict the label using a non-linear hypothesis map $h^{(\mathbf{w})}(x) \in \mathcal{H}_{\text{Gauss}}^{(2)}$ with some weight vector $\mathbf{w} \in \mathbb{R}^2$

map Φ. We can extend linear regression using an arbitrary feature map

$$\Phi(x) = (\phi_1(x), \ldots, \phi_n(x))^T \tag{3.11}$$

with the scalar maps $\phi_j : \mathbb{R} \to \mathbb{R}$ which are referred to as **basis functions**. The choice of basis functions depends heavily on the particular application and the underlying relation between features and labels of the observed datapoints. The basis functions underlying polynomial regression are $\phi_j(x) = x^j$.

Another popular choice for the basis functions are "Gaussians"

$$\phi_{\sigma,\mu}(x) = \exp(-(1/(2\sigma^2))(x-\mu)^2). \tag{3.12}$$

The family (3.12) of maps is parameterized by the variance σ^2 and the mean (shift) μ. We obtain **Gaussian basis linear regression** by combining the feature map

$$\Phi(x) = \left(\phi_{\sigma_1,\mu_1}(x), \ldots, \phi_{\sigma_n,\mu_n}(x)\right)^T \tag{3.13}$$

with linear regression (see Fig. 3.5). The resulting hypothesis space is then

$$\mathcal{H}_{\text{Gauss}}^{(n)} = \{h^{(\mathbf{w})} : \mathbb{R} \to \mathbb{R} : h^{(\mathbf{w})}(x) = \sum_{j=1}^{n} w_j \phi_{\sigma_j,\mu_j}(x)$$

$$\text{with weights } \mathbf{w} = (w_1, \ldots, w_n)^T \in \mathbb{R}^n \}. \tag{3.14}$$

Different choices for the variance σ^2 and shifts μ_j of the Gaussian function in (3.12) results in different hypothesis spaces $\mathcal{H}_{\text{Gauss}}$. Section 6.3 will discuss model selection techniques that allow to find useful values for these parameters.

The hypotheses of (3.14) are parameterized by a weight vector $\mathbf{w} \in \mathbb{R}^n$. Each hypothesis in $\mathcal{H}_{\text{Gauss}}$ corresponds to a particular choice for the weight vector $\mathbf{w}$. Thus, instead of searching over $\mathcal{H}_{\text{Gauss}}$ to find a good hypothesis, we can search over $\mathbb{R}^n$.

3.6 Logistic Regression

Logistic regression is a method for classifying datapoints which are characterized by feature vectors $\mathbf{x} \in \mathbb{R}^n$ (feature space $\mathcal{X} = \mathbb{R}^n$) according to two categories which are encoded by a label y. It will be convenient to use the label space $\mathcal{Y} = \mathbb{R}$ and encode the two label values as $y = 1$ and $y = -1$. Logistic regression learns a hypothesis out of the hypothesis space $\mathcal{H}^{(n)}$ (see (3.1)).[2] Note that the hypothesis space is the same as used in linear regression (see Sect. 3.1).

At first sight, it seems wasteful to use a linear hypothesis $h(\mathbf{x}) = \mathbf{w}^T\mathbf{x}$, with some weight vector $\mathbf{w} \in \mathbb{R}^n$, to predict a binary label y. Indeed, while the prediction $h(\mathbf{x})$ can take any real number, the label $y \in \{-1, 1\}$ takes on only one of the two real numbers 1 and -1.

It turns out that even for binary labels it is quite useful to use a hypothesis map h which can take on arbitrary real numbers. We can always obtain a predicted label $\hat{y} \in \{-1, 1\}$ by comparing hypothesis value $h(\mathbf{x})$ with a threshold. A data point with features $\mathbf{x}$, is classified as $\hat{y} = 1$ if $h(\mathbf{x}) \geq 0$ and $\hat{y} = -1$ for $h(\mathbf{x}) < 0$. Thus, we use the sign of the predictor h to determine the final prediction for the label. The absolute value $|h(\mathbf{x})|$ is then used to quantify the reliability of (or confidence in) the classification $\hat{y}$.

Consider two datapoints with feature vectors $\mathbf{x}^{(1)}, \mathbf{x}^{(2)}$ and a linear classifier map h yielding the function values $h(\mathbf{x}^{(1)}) = 1/10$ and $h(\mathbf{x}^{(2)}) = 100$. Whereas the predictions for both datapoints result in the same label predictions, i.e., $\hat{y}^{(1)} = \hat{y}^{(2)} = 1$, the classification of the data point with feature vector $\mathbf{x}^{(2)}$ seems to be much more reliable.

Logistic regression uses the logistic loss (2.12) to assess the quality of a particular hypothesis $h^{(\mathbf{w})} \in \mathcal{H}^{(n)}$. In particular, given some labeled training set $\mathcal{D} = \{\mathbf{x}^{(i)}, y^{(i)}\}_{i=1}^m$, logistic regression tries to minimize the empirical risk (average logistic loss)

$$\widehat{L}(\mathbf{w}|\mathcal{D}) = (1/m) \sum_{i=1}^m \log(1 + \exp(-y^{(i)} h^{(\mathbf{w})}(\mathbf{x}^{(i)})))$$

$$\overset{h^{(\mathbf{w})}(\mathbf{x})=\mathbf{w}^T\mathbf{x}}{=} (1/m) \sum_{i=1}^m \log(1 + \exp(-y^{(i)} \mathbf{w}^T \mathbf{x}^{(i)})). \qquad (3.15)$$

Once we have found the optimal weight vector $\widehat{\mathbf{w}}$ which minimizes (3.15), we classify a datapoint based on its features $\mathbf{x}$ according to

$$\hat{y} = \begin{cases} 1 & \text{if } h^{(\widehat{\mathbf{w}})}(\mathbf{x}) \geq 0 \\ -1 & \text{otherwise.} \end{cases} \qquad (3.16)$$

---

[2] It is important to note that logistic regression can be used with an arbitrary label space which contains two different elements. Another popular choice for the label space is $\mathcal{Y} = \{0, 1\}$.

Since $h^{(\widehat{\mathbf{w}})}(\mathbf{x}) = (\widehat{\mathbf{w}})^T \mathbf{x}$ (see (3.1)), the classifier (3.16) amounts to testing whether $(\widehat{\mathbf{w}})^T \mathbf{x} \geq 0$ or not.

The classifier (3.16) partitions the feature space $\mathcal{X} = \mathbb{R}^n$ into two half-spaces $\mathcal{R}_1 = \{\mathbf{x} : (\widehat{\mathbf{w}})^T \mathbf{x} \geq 0\}$ and $\mathcal{R}_{-1} = \{\mathbf{x} : (\widehat{\mathbf{w}})^T \mathbf{x} < 0\}$ which are separated by the hyperplane $(\widehat{\mathbf{w}})^T \mathbf{x} = 0$ (see Fig. 2.9). Any datapoint with features $\mathbf{x} \in \mathcal{R}_1$ ($\mathbf{x} \in \mathcal{R}_{-1}$) is classified as $\hat{y} = 1$ ($\hat{y} = -1$).

Logistic regression can be interpreted as a maximum likelihood estimator within a particular probabilistic model for the datapoints. This probabilistic model interprets the label $y \in \{-1, 1\}$ of a datapoint as a RV with the probability distribution

$$p(y = 1; \mathbf{w}) = 1/(1 + \exp(-\mathbf{w}^T \mathbf{x}))$$

$$\overset{h^{(\mathbf{w})}(\mathbf{x}) = \mathbf{w}^T \mathbf{x}}{=} 1/(1 + \exp(-h^{(\mathbf{w})}(\mathbf{x})))). \tag{3.17}$$

As the notation indicates, the probability (3.17) is parameterized by the weight vector $\mathbf{w}$ of the linear hypothesis $h^{(\mathbf{w})}(\mathbf{x}) = \mathbf{w}^T \mathbf{x}$. Given the probabilistic model (3.17), we can interpret the classification (3.16) as choosing $\hat{y}$ to maximize the probability $p(y = \hat{y}; \mathbf{w})$.

Since $p(y = 1) + p(y = -1) = 1$,

$$p(y = -1) = 1 - p(y = 1)$$

$$\overset{(3.17)}{=} 1 - 1/(1 + \exp(-\mathbf{w}^T \mathbf{x}))$$

$$= 1/(1 + \exp(\mathbf{w}^T \mathbf{x})). \tag{3.18}$$

In practice we do not know the weight vector in (3.17). Rather, we have to estimate the weight vector $\mathbf{w}$ in (3.17) from observed datapoints. A principled approach to estimate the weight vector is to maximize the probability (or likelihood) of actually obtaining the dataset $\mathcal{D} = \{(\mathbf{x}^{(i)}, y^{(i)})\}_{i=1}^m$ as realizations of i.i.d. data points whose labels are distributed according to (3.17). This yields the maximum likelihood estimator

$$\widehat{\mathbf{w}} = \underset{\mathbf{w} \in \mathbb{R}^n}{\text{argmax}}\, p(\{y^{(i)}\}_{i=1}^m)$$

$$\overset{y^{(i)} \text{i.i.d.}}{=} \underset{\mathbf{w} \in \mathbb{R}^n}{\text{argmax}} \prod_{i=1}^m p(y^{(i)})$$

$$\overset{(3.17),(3.18)}{=} \underset{\mathbf{w} \in \mathbb{R}^n}{\text{argmax}} \prod_{i=1}^m 1/(1 + \exp(-y^{(i)} \mathbf{w}^T \mathbf{x}^{(i)})). \tag{3.19}$$

Note that the last expression (3.19) is only valid if we encode the binary labels using the values 1 and -1. Using different label values results in a different expression.

Maximizing a positive function $f(\mathbf{w}) > 0$ is equivalent to maximizing $\log f(x)$,

$$\operatorname*{argmax}_{\mathbf{w}\in\mathbb{R}^n} f(\mathbf{w}) = \operatorname*{argmax}_{\mathbf{w}\in\mathbb{R}^n} \log f(\mathbf{w}).$$

Therefore, (3.19) can be further developed as

$$\widehat{\mathbf{w}} \overset{(3.19)}{=} \operatorname*{argmax}_{\mathbf{w}\in\mathbb{R}^n} \sum_{i=1}^{m} -\log\left(1+\exp(-y^{(i)}\mathbf{w}^T\mathbf{x}^{(i)})\right)$$

$$= \operatorname*{argmin}_{\mathbf{w}\in\mathbb{R}^n}(1/m) \sum_{i=1}^{m} \log\left(1+\exp(-y^{(i)}\mathbf{w}^T\mathbf{x}^{(i)})\right). \tag{3.20}$$

Comparing (3.20) with (3.15) reveals that logistic regression is nothing but maximum likelihood estimation of the weight vector $\mathbf{w}$ in the probabilistic model (3.17).

3.7 Support Vector Machines

Support vector machines are a family of ML methods for learning a hypothesis to predict a binary label y of a data point based on its features $\mathbf{x}$. Without loss of generality we consider binary labels taking values in the label space $\mathcal{Y} = \{-1, 1\}$. A support vector machine uses the linear hypothesis space (3.1) which consists of linear maps $h(\mathbf{x}) = \mathbf{w}^T\mathbf{x}$ with some weight vector $\mathbf{w} \in \mathbb{R}^n$. Thus, the support vector machine uses the same hypothesis space as linear regression and logistic regression which we have discussed in Sects. 3.1 and 3.6, respectively. What sets the support vector machine apart from these other methods is the choice of loss function.

Different instances of a support vector machine are obtained by using different constructions for the features of a data point. Kernel support vector machines use the concept of a kernel map to construct (typically high-dimensional) features (see Sect. 3.9 and [5]). In what follows, we assume the feature construction has been solved and we have access to a feature vector $\mathbf{x} \in \mathbb{R}^n$ for each data point.

Figure 3.6 depicts a dataset $\mathcal{D}$ of labeled data points, each characterized by a feature vector $\mathbf{x}^{(i)} \in \mathbb{R}^2$ (used as coordinates of a marker) and a binary label $y^{(i)} \in \{-1, 1\}$ (indicated by different marker shapes). We can partition dataset $\mathcal{D}$ into two classes

$$\mathcal{C}^{(y=1)} = \{\mathbf{x}^{(i)} : y^{(i)} = 1\}, \text{ and } \mathcal{C}^{(y=-1)} = \{\mathbf{x}^{(i)} : y^{(i)} = -1\}. \tag{3.21}$$

The support vector machine tries to learn a linear map $h^{(\mathbf{w})}(\mathbf{x}) = \mathbf{w}^T\mathbf{x}$ that perfectly separates the two classes in the sense of

$$\underbrace{h\left(\mathbf{x}^{(i)}\right)}_{\mathbf{w}^T\mathbf{x}^{(i)}} > 0 \text{ for } \mathbf{x}^{(i)} \in \mathcal{C}^{(y=1)} \text{ and } \underbrace{h\left(\mathbf{x}^{(i)}\right)}_{\mathbf{w}^T\mathbf{x}^{(i)}} < 0 \text{ for } \mathbf{x}^{(i)} \in \mathcal{C}^{(y=-1)}. \tag{3.22}$$

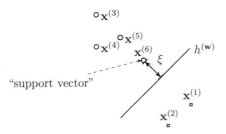

Fig. 3.6 The support vector machine learns a hypothesis (or classifier) $h^{(\mathbf{w})}$ with minimum average soft-margin hinge loss (3.23). Minimizing this loss is equivalent to maximizing the margin ξ between the decision boundary of $h^{(\mathbf{w})}$ and each class of the training set

We refer to a dataset, whose datapoints have binary labels. as linear separable if we can find at least one linear map that separates in the sense of (3.22). The dataset in Fig. 3.6 is in early separable.

As can be verified easily, any linear map $h^{(\mathbf{w})}(\mathbf{x}) = \mathbf{w}^T\mathbf{x}$ achieving zero average hinge loss (2.11) on the dataset $\mathcal{D}$ perfectly satisfies this dataset (3.22). It seems reasonable to learn a linear map by minimizing the average hinge loss (2.11). However, one drawback of this approach is that there might be (infinitely) many different linear maps that achieve zero average hinge loss and, in turn, perfectly separate the data points in Fig. 3.6. Indeed, consider a linear map $h^{(\mathbf{w})}$ that achieves zero average hinge loss for the $\mathcal{D}$ in Fig. 3.6 (and therefore perfectly separates it). Then, any other linear map $h^{(\mathbf{w}')}$ with weights $\mathbf{w}' = \lambda\mathbf{w}$, using an arbitrary number $\lambda > 1$ also achieves zero average hinge loss (and perfectly separates the dataset).

Neither the separability requirement (3.22) nor the hinge loss (2.11) are sufficient as a sole training criterion. Indeed, there are many (if not most) datasets that are not linearly separable. Even for a linearly separable dataset (such as the one Figure 3.6), there are infinitely many linear maps with zero average hinge loss. Which one of these infinitely many different maps should we use? To settle these issues, the support vector machine uses a "regularized" hinge loss,

$$L((\mathbf{x}, y), h^{(\mathbf{w})}) := \max\{0, 1 - y \cdot h^{(\mathbf{w})}(\mathbf{x})\} + \lambda\|\mathbf{w}\|^2$$

$$\overset{h^{(\mathbf{w})}(\mathbf{x})=\mathbf{w}^T\mathbf{x}}{=} \max\{0, 1 - y \cdot \mathbf{w}^T\mathbf{x}\} + \lambda\|\mathbf{w}\|^2. \qquad (3.23)$$

The loss (3.23) augments the hinge loss (2.11) by the term $\lambda\|\mathbf{w}\|^2$. This term is the scaled (by $\lambda > 0$) squared Euclidean norm of the weights $\mathbf{w}$ of the linear hypothesis h used to classify data points. it can be shown that adding the term $\lambda\|\mathbf{w}\|^2$ to the hinge loss (2.11) has an regularization effect. Loosely speaking, the resulting loss favours linear maps $h^{(\mathbf{w})}$ that are robust against (small) perturbations of the data points. The tuning parameter λ in (3.23) controls the strength of this regularization effect and might therefore also be referred to as a regularization parameter. We will discuss the basic principles of regularization on a more general level in Chap. 7.

Let us now develop a useful geometric interpretation of the linear hypothesis obtained by minimizing the loss function (3.23). According to [5, Chap. 2], a classifier $h^{(\mathbf{w}_{\text{SVM}})}$ that minimizes the average loss (3.23), maximizes the distance (margin) ξ between its decision boundary and each of the two classes $\mathcal{C}^{(y=1)}$ and $\mathcal{C}^{(y=-1)}$ (see (3.21)). The decision boundary is given by the set of feature vectors $\mathbf{x}$ satisfying $\mathbf{w}_{\text{SVM}}^{T}\mathbf{x} = 0$,

Making the margin as large as possible is reasonable as it ensures that the resulting classifications are robust against small perturbations of the features (see Sect. 7.2). As depicted in Fig. 3.6, the margin between the decision boundary and the classes $\mathcal{C}_1$ and $\mathcal{C}_2$ is typically determined by few datapoints (such as $\mathbf{x}^{(6)}$ in Fig. 3.6) which are closest to the decision boundary. These data points have minimum distance to the decision boundary and are referred to as support vectors.

We highlight that both, the support vector machine and logistic regression use the same hypothesis space of linear maps. Therefore, both methods learn a linear classifier $h^{(\mathbf{w})} \in \mathcal{H}^{(n)}$ (see (3.1)) whose decision boundary is a hyperplane in the feature space $\mathcal{X} = \mathbb{R}^n$ (see Fig. 2.9). The difference between support vector machine and logistic regression is in their choice for the loss function used to evaluate the quality of a hypothesis $h^{(\mathbf{w})} \in \mathcal{H}^{(n)}$.

The hinge loss (2.11) is (in some sense) the best convex approximation to the $0/1$ loss (2.9). Thus, we expect the classifier obtained by the support vector machine to yield a smaller classification error probability $p(\hat{y} \neq y)$ (with $\hat{y} = 1$ if $h(\mathbf{x}) \geq 0$ and $\hat{y} = -1$ otherwise) compared to logistic regression which uses the logistic loss (2.12). The support vector machine is also statistically appealing as it learns a robust hypothesis. Indeed, learning the hypothesis with maximum margin implies that the resulting classifier is maximally robust against perturbations of the feature vectors of data points. Section 7.2 discusses the importance of robustness in ML methods in more detail.

The statistical superiority of the support vector machine comes at the cost of increased computational complexity. In particular, the hinge loss (2.11) is non-differentiable which prevents the use of simple gradient-based methods (see Chap. 5) and requires more advanced optimization methods. In contrast, the logistic loss (2.12) is convex and differentiable. We can therefore use gradient based methods to minimize the average logistic loss incurred on a training set (see Chap. 5).

3.8 Bayes Classifier

Consider datapoints characterized by features $\mathbf{x} \in \mathcal{X}$ and some binary label $y \in \mathcal{Y}$. We can use any two different label values but let us assume that the two possible label values are $y = -1$ or $y = 1$. We would like to find (or learn) a classifier $h : \mathcal{X} \to \mathcal{Y}$ such that the predicted (or estimated) label $\hat{y} = h(\mathbf{x})$ agrees with the true label $y \in \mathcal{Y}$ as much as possible. Thus, it is reasonable to assess the quality of a classifier h using the $0/1$ loss (2.9). We could then learn a classifier using the empirical risk minimization with the loss function (2.9). However, the resulting optimization problem is typically intractable since the loss (2.9) is non-convex and non-differentiable.

Instead of solving the (intractable) empirical risk minimization for $0/1$ loss, we take a different route to construct a classifier. This construction is based on a simple probabilistic model for the datapoints. Using this model, we can interpret the average $0/1$ loss on training data as an approximation for the probability $P_{err} = p(y \neq h(\mathbf{x}))$. Any classifier that minimizes this error probability is referred to as a Bayes estimator. Note that the Bayes estimator depends on the probabilistic model for the data points. We obtain different glsbayesestimators for different probabilistic models.

One widely used probabilistic model results in a Bayes estimator that belongs to the linear hypothesis space (3.1). Note that this hypothesis space underlies also logistic regression (see Sect. 3.6) and the support vector machine (see Sect. 3.7). Thus, logistic regression, support vector machine and Bayes estimator are all examples of a linear classifier (see Fig. 2.9).

A linear classifier partitions the feature space $\mathcal{X}$ into two half-spaces. One half-space consists of all feature vectors $\mathbf{x}$ which result in the predicted label $\hat{y} = 1$ and the other half-space constituted by all feature vectors $\mathbf{x}$ which result in the predicted label $\hat{y} = -1$. The family of ML methods that learn a linear classifier differ in their choices for the loss function and, in turn, how they choose these half-spaces. Section 4.5 will discuss ML methods using Bayes estimator in more detail.

3.9 Kernel Methods

Consider a ML (classification or regression) problem with an underlying feature space $\mathcal{X}$. In order to predict the label $y \in \mathcal{Y}$ of a datapoint based on its features $\mathbf{x} \in \mathcal{X}$, we apply a predictor h selected out of some hypothesis space $\mathcal{H}$. Let us assume that the available computational infrastructure only allows us to use a linear hypothesis space $\mathcal{H}^{(n)}$ (see (3.1)).

For some applications, using a linear hypothesis $h(\mathbf{x}) = \mathbf{w}^T \mathbf{x}$ is not suitable since the relation between features $\mathbf{x}$ and label y might be highly non-linear. One approach to extend the capabilities of linear hypotheses is to transform the raw features of a data point before applying a linear hypothesis h.

The family of kernel methods is based on transforming the features $\mathbf{x}$ to new features $\hat{\mathbf{x}} \in \mathcal{X}'$ which belong to a (typically very) high-dimensional space $\mathcal{X}'$ [5]. It is not uncommon that, while the original feature space is a low-dimensional Euclidean space (e.g., $\mathcal{X} = \mathbb{R}^2$), the transformed feature space $\mathcal{X}'$ is an infinite-dimensional function space.

The rationale behind transforming the original features into a new (higher-dimensional) feature space $\mathcal{X}'$ is to reshape the intrinsic geometry of the feature vectors $\mathbf{x}^{(i)} \in \mathcal{X}$ such that the transformed feature vectors $\hat{\mathbf{x}}^{(i)}$ have a "simpler" geometry (see Fig. 3.7).

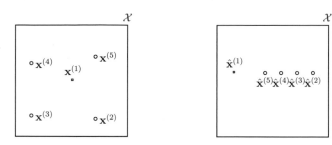

Fig. 3.7 The data set $\mathcal{D} = \{(\mathbf{x}^{(i)}, y^{(i)})\}_{i=1}^{5}$ consists of 5 datapoints with features $\mathbf{x}^{(i)}$ and binary labels $y^{(i)}$. Left: In the original feature space $\mathcal{X}$, the datapoints cannot be separated perfectly by any linear classifier. Right: The feature map $\phi : \mathcal{X} \to \mathcal{X}'$ transforms the features $\mathbf{x}^{(i)}$ to the new features $\hat{\mathbf{x}}^{(i)} = \phi(\mathbf{x}^{(i)})$ in the new feature space $\mathcal{X}'$. In the new feature space $\mathcal{X}'$ the datapoints can be separated perfectly by a linear classifier

Kernel methods are obtained by formulating ML problems (such as linear regression or logistic regression) using the transformed features $\hat{\mathbf{x}} = \phi(\mathbf{x})$. A key challenge within kernel methods is the choice of the feature map $\phi : \mathcal{X} \to \mathcal{X}'$ which maps the original feature vector $\mathbf{x}$ to a new feature vector $\hat{\mathbf{x}} = \phi(\mathbf{x})$.

3.10   Decision Trees

A decision tree is a flowchart-like description of a map $h : \mathcal{X} \to \mathcal{Y}$ which maps the features $\mathbf{x} \in \mathcal{X}$ of a datapoint to a predicted label $h(\mathbf{x}) \in \mathcal{Y}$ [6]. While decision trees can be used for arbitrary feature space $\mathcal{X}$ and label space $\mathcal{Y}$, we will discuss them for the particular feature space $\mathcal{X} = \mathbb{R}^2$ and label space $\mathcal{Y} = \mathbb{R}$.

Figure 3.8 depicts an example for a decision tree. A decision tree consists of nodes which are connected by directed edges. We can think of a decision tree as a step-by-step instruction, or a "recipe", for how to compute the function value $h(\mathbf{x})$ given the features $\mathbf{x} \in \mathcal{X}$ of a datapoint. This computation starts at the **root node** and ends at one of the **leaf nodes** of the decision tree.

A leaf node m, which does not have any outgoing edges, represents a decision region $\mathcal{R}_m \subseteq \mathcal{X}$ in the feature space. The hypothesis h associated with a decision tree is constant over the regions $\mathcal{R}_m$, such that $h(\mathbf{x}) = h_m$ for all $\mathbf{x} \in \mathcal{R}_m$ and some fixed number $h_m \in \mathbb{R}$.

In general, there are two types of nodes in a decision tree:

- decision (or test) nodes, which represent particular "tests" about the feature vector $\mathbf{x}$ (e.g., "is the norm of $\mathbf{x}$ larger than 10?").
- leaf nodes, which correspond to subsets of the feature space.

The particular decision tree depicted in Fig. 3.8 consists of two decision nodes (including the root node) and three leaf nodes.

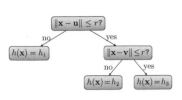

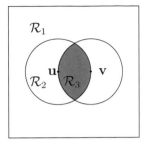

Fig. 3.8 A decision tree represents a hypothesis h which is constant on subsets $\mathcal{R}_m$, i.e., $h(\mathbf{x}) = h_m$ for all $\mathbf{x} \in \mathcal{R}_m$. Each subset $\mathcal{R}_m \subseteq \mathcal{X}$ corresponds to a leaf node in the decision tree

Fig. 3.9 A hypothesis space $\mathcal{H}$ consisting of two decision trees with depth at most 2 and using the tests $\|\mathbf{x} - \mathbf{u}\| \le r$ and $\|\mathbf{x} - \mathbf{v}\| \le r$ with a fixed radius r and vectors $\mathbf{u}, \mathbf{v} \in \mathbb{R}^n$

Given limited computational resources, we can only use decision trees which are not too deep. Consider the hypothesis space consisting of all decision trees which use the tests "$\|\mathbf{x} - \mathbf{u}\| \le r$" and "$\|\mathbf{x} - \mathbf{v}\| \le r$" , with some vectors $\mathbf{u}$ and $\mathbf{v}$, some positive radius $r > 0$ and depth no larger than 2.[3]

To assess the quality of a particular decision tree we can use various loss functions. Examples of loss functions used to measure the quality of a decision tree are the squared error loss (for numeric labels) or the impurity of individual decision regressions (for discrete labels).

Decision tree methods use as a hypothesis space the set of all hypotheses which represented by some collection of decision trees. Figure 3.9 depicts a collection of decision trees which are characterized by having depth at most two. These methods search for a decision trees such that the corresponding hypothesis has minimum average loss on some labeled training data (see Sect. 4.4).

A collection of decision trees can be constructed based on a fixed set of "elementary tests" on the input feature vector, e.g., $\|\mathbf{x}\| > 3, x_3 < 1$ or a continuous ensemble of parametrized tests such as $\{x_2 > \eta\}_{\eta \in [0, 10]}$. We then build a hypothesis space by considering all decision trees not exceeding a maximum depth and whose decision nodes carry out one of the elementary tests.

A decision tree represents a map $h : \mathcal{X} \to \mathcal{Y}$, which is piecewise-constant over regions of the feature space $\mathcal{X}$. These non-overlapping regions form a partitioning

---

[3] The depth of a decision tree is the maximum number of hops it takes to reach a leaf node starting from the root and following the arrows. The decision tree depicted in Fig. 3.8 has depth 2.

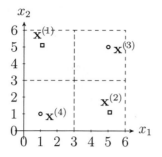

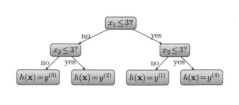

Fig. 3.10 Using a sufficiently large (deep) decision tree, we can construct a map h that perfectly fits any given labeled dataset $\{(\mathbf{x}^{(i)}, y^{(i)})\}_{i=1}^{m}$ such that $h(\mathbf{x}^{(i)}) = y^{(i)}$ for $i = 1, \ldots, m$

of the feature space. Each leaf node of a decision tree corresponds to one particular region. Using large decision trees, which involve many different test nodes, we can represent very complicated partitions that resemble any given labeled dataset (see Fig. 3.10).

This is quite different from ML methods using the linear hypothesis space (3.1), such as linear regression, logistic regression or the support vector machine. These methods learn linear hypothesis maps with a rather simple geometry. Indeed, a linear map is constant along hyperplanes. Moreover, the decision regions obtained from linear classifiers are always entire half-spaces (see Fig. 2.9).

In contrast, the shape of a map represented by a decision tree can be much more complicated. Using a sufficiently large (deep) decision tree, we can obtain a hypothesis map that closely approximates any given non-linear map. Using sufficiently deep decision trees for classification problems allows for highly irregular decision regions.

3.11   Deep Learning

Another example of a hypothesis space uses a signal-flow representation of a hypothesis map $h : \mathbb{R}^n \rightarrow \mathbb{R}$. This signal-flow representation is referred to as artificial neural network. Figure 3.8 depicts an example for a artificial neural network that is used to represent a (parameterized) hypothesis $h^{(\mathbf{w})} : \mathbb{R}^n \rightarrow \mathbb{R}$. A feature vector $\mathbf{x} \in \mathbb{R}^n$ is fed into the input units, each of which reads in one single feature $x_j \in \mathbb{R}$. The features x_j are then multiplied with the weights $w_{j,j'}$ associated with the link between the jth input node ("neuron") with the j'th node in the middle (hidden) layer. The output of the j'-th node in the hidden layer is given by $s_{j'} = g(\sum_{j=1}^{n} w_{j,j'} x_j)$ with some (typically non-linear) activation function $f : \mathbb{R} \rightarrow \mathbb{R}$. The input argument to the activation function is the weighted combination $\sum_{j=1}^{n} w_{j,j'} s_{j'}$ of the outputs s_j of the nodes in a previous layer. For the artificial neural network depicted in Fig. 3.11, the output of neuron s_1 is $f(z)$ with $z = w_{1,1} x_1 + w_{1,2} x_2$.

Fig. 3.11 Artificial neural network representation of a predictor $h^{(\mathbf{w})}(\mathbf{x})$ which maps the input (feature) vector $\mathbf{x} = (x_1, x_2)^T$ to a predicted label (output) $h^{(\mathbf{w})}(\mathbf{x})$

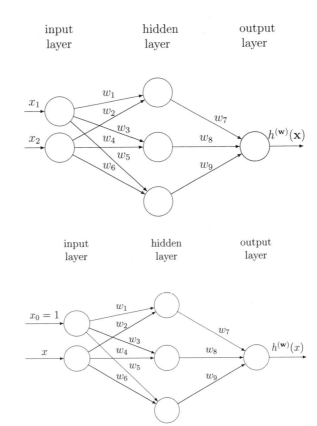

Fig. 3.12 An artificial neural network with one hidden layer defines a hypothesis space consisting of all maps $h^{(\mathbf{w})}(x)$ obtained from all possible choices for the weights $\mathbf{w} = (w_1, \ldots, w_9)^T$

Two popular choices for the activation function used within artificial neural networks are the sigmoid function $f(z) = \frac{1}{1+\exp(-z)}$ or the deep net $f(z) = \max\{0, z\}$. Artificial neural networks using many, say 10, hidden layers, is often referred to as a deep net. ML methods using hypothesis spaces obtained from deep nets are known as deep learning methods [7].

Remarkably, using some simple non-linear activation function $f(z)$ as the building block for artificial neural networks allows us to represent an extremely large class of predictor maps $h^{(\mathbf{w})} : \mathbb{R}^n \to \mathbb{R}$. The hypothesis space generated by a given artificial neural network structure, i.e., the set of all predictor maps which can be implemented by a given artificial neural network and suitable weights $\mathbf{w}$, tends to be much larger than the hypothesis space (2.4) of linear predictors using weight vectors $\mathbf{w}$ of the same length [7, Chap. 6.4.1.]. It can be shown that an artificial neural network with only one single (but arbitrarily large) hidden layer can approximate any given map $h : \mathcal{X} \to \mathcal{Y} = \mathbb{R}$ to any desired accuracy [8]. However, a key insight which underlies many deep learning methods is that using several layers with few neurons, instead of one single layer containing many neurons, is computationally favourable [9].

Fig. 3.13 Each single
neuron of the artificial neural
network depicted in Figure
3.12 implements a weighted
summation $z = \sum_j w_j x_j$ of
its inputs x_j followed by
applying a non-linear
activation function $f(z)$

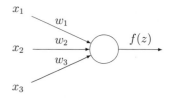

The recent success of ML methods based on artificial neural network with many
hidden layers (which makes them deep) might be attributed to the fact that the network
representation of hypothesis maps is beneficial for the computational implementation of ML methods. First, we can evaluate a map $h^{(\mathbf{w})}$ represented by an artificial
neural network efficiently using modern parallel and distributed computing infrastructure via message passing over the network. Second, the graphical representation
of a parametrized hypothesis in the form of a artificial neural network allows us to
efficiently compute the gradient of the loss function via a (highly scalable) message
passing procedure known as back-propagation [7].

3.12 Maximum Likelihood

For many applications it is useful to model the observed datapoints $\mathbf{z}^{(i)}$, with $i =
1, \ldots, m$, as i.i.d. realizations of a random variable $\mathbf{z}$ with probability distribution
$p(\mathbf{z}; \mathbf{w})$. This probability distribution is parameterized in the sense of depending on
a weight vector $\mathbf{w} \in \mathbb{R}^n$. A principled approach to estimating the vector $\mathbf{w}$ based on a
set of i.i.d. realizations $\mathbf{z}^{(1)}, \ldots, \mathbf{z}^{(m)} \sim p(\mathbf{z}; \mathbf{w})$ is **maximum likelihood estimation**
[10].

Maximum likelihood estimation can be interpreted as an ML problem with a
hypothesis space parameterized by the weight vector $\mathbf{w}$, i.e., each element $h^{(\mathbf{w})}$ of
the hypothesis space $\mathcal{H}$ corresponds to one particular choice for the weight vector
$\mathbf{w}$, and the loss function

$$L(\mathbf{z}, h^{(\mathbf{w})}) := - \log p(\mathbf{z}; \mathbf{w}). \tag{3.24}$$

A widely used choice for the probability distribution $p(\mathbf{z}; \mathbf{w})$ is a multivariate
normal (Gaussian) distribution with mean μ and covariance matrix Σ, both of which
constitute the weight vector $\mathbf{w} = (\mu, \Sigma)$ (we have to reshape the matrix Σ suitably
into a vector form). Given the i.i.d. realizations $\mathbf{z}^{(1)}, \ldots, \mathbf{z}^{(m)} \sim p(\mathbf{z}; \mathbf{w})$, the maximum likelihood estimates $\hat{\mu}, \widehat{\Sigma}$ of the mean vector and the covariance matrix are
obtained via

$$\hat{\mu}, \widehat{\Sigma} = \underset{\mu \in \mathbb{R}^n, \Sigma \in \mathbb{S}^n_+}{\operatorname{argmin}} (1/m) \sum_{i=1}^{m} - \log p\big(\mathbf{z}^{(i)}; (\mu, \Sigma)\big). \tag{3.25}$$

The optimization in (3.25) is over all possible choices for the mean vector $\mu \in \mathbb{R}^n$ and the covariance matrix $\Sigma \in \mathbb{S}_+^n$. Here, $\mathbb{S}_+^n$ denotes the set of all positive semi-definite Hermitian $n \times n$ matrices.

The maximum likelihood problem (3.25) can be interpreted as an instance of empirical risk minimization (4.3) using the particular loss function (3.24). The resulting estimates are given explicitly as

$$\hat{\mu} = (1/m) \sum_{i=1}^{m} \mathbf{z}^{(i)}, \text{ and } \widehat{\Sigma} = (1/m) \sum_{i=1}^{m} (\mathbf{z}^{(i)} - \hat{\mu})(\mathbf{z}^{(i)} - \hat{\mu})^T. \qquad (3.26)$$

Note that the expressions (3.26) are valid only when the probability distribution of the datapoints is modelled as a multivariate normal distribution.

3.13   Nearest Neighbour Methods

Nearest neighbour methods are an important family of ML methods that are characterized by a specific construction of the hypothesis space. This family provides methods for regression problems involving numeric labels (e.g., with label space $\mathcal{Y} = \mathbb{R}$) as well as for classification problems involving categorical labels (e.g., with label space $\mathcal{Y} = \{-1, 1\}$). While nearest neighbour methods can be combined with arbitrary label spaces, they require the feature space to be a metric space [1] so we can compute distances between different feature vectors.

A widely used example for a metric feature space is the Euclidean space $\mathbb{R}^n$ with the Euclidean distance $\|\mathbf{x} - \mathbf{x}'\|$ between two vectors $\mathbf{x}, \mathbf{x}' \in \mathbb{R}^n$. Consider a dataset $\mathcal{D} = \{(\mathbf{x}^{(i)}, y^{(i)})\}_{i=1}^{m}$ of labeled data points, each one characterized by a feature vector and a label. Nearest neighbour methods use a hypothesis space that consist of piece-wise maps $h : \mathcal{X} \to \mathcal{Y}$. The function value $h(\mathbf{x})$, for some feature vector $\mathbf{x}$, depends only on the (labels of the) k nearest data points in the dataset $\mathcal{D}$. The number k of nearest neighbours is a design parameter of the method. Nearest neighbour methods are also referred to as k-nearest neighbour (k-NN) methods to make their dependence on the parameter k explicit.

It is important to note that, in contrast to the ML methods in Sects. 3.1–3.11, the hypothesis space of k-NN depends on a (training) dataset $\mathcal{D}$. As a consequence, k-NN methods need to query (read in) the training set whenever the compute a prediction. In particular, to compute a prediction $h(\mathbf{x})$ for a new data point with features $\mathbf{x}$, k-NN needs to determine the nearest neighbours in the training set. When using a large training set (which is typically beneficial for the resulting accuracy of the ML method) this implies a large storage requirement for k-NN methods. Moreover, k-NN methods might be prone to revealing sensitive information with its predictions (see Exercise 3.7).

Fig. 3.14 A hypothesis map h for k-NN with $k = 1$ and feature space $\mathcal{X} = \mathbb{R}^2$. The hypothesis map is constant over regions (indicated by the coloured areas) located around feature vectors $\mathbf{x}^{(i)}$ (indicated by a dot) of a dataset $\mathcal{D} = \{(\mathbf{x}^{(i)}, y^{(i)})\}$

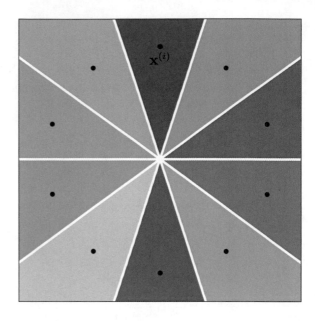

3.14 Deep Reinforcement Learning

Deep reinforcement learning (DRL) refers to a subset of ML problems and methods that revolve around the control of dynamic systems such as autonomous driving cars or cleaning robots [11–13]. A DRL problem involves data points that represent the states of a dynamic system at different time instants $t = 0, 1, \ldots$. The data points representing the state at some time instant t is characterized by the feature vector $\mathbf{x}^{(t)}$. The entries of this feature vector are the individual features of the state at time t. These features might be obtained via sensors, onboard-cameras or other ML methods (that predict the location of the dynamic system). The label $y^{(t)}$ of a data point might represent the optimal steering angle at time t.

DRL methods learn a hypothesis h that delivers optimal predictions $\hat{y}^{(t)} := h(\mathbf{x}^{(t)})$ for the optimal steering angle $y^{(t)}$. As their name indicates, DRL methods use hypothesis spaces obtained from a deep net (see Sect. 3.11). The quality of the prediction $\hat{y}^{(t)}$ obtained from a hypothesis is measured by the loss $L((\mathbf{x}^{(t)}, y^{(t)}), h) := -r^{(t)}$ with a reward signal $r^{(t)}$. This reward signal might be obtained from a distance (collision avoidance) sensor or low-level characteristics of an on-board camera snapshot.

The (negative) reward signal $-r^{(t)}$ typically depends on the feature vector $\mathbf{x}^{(t)}$ and the discrepancy between optimal steering direction $y^{(t)}$ (which is unknown) and its prediction $\hat{y}^{(t)} := h(\mathbf{x}^{(t)})$. However, what sets DRL methods apart from other ML methods such as linear regression (see Sect. 3.1) or logistic regression (see Sect. 3.6) is that they can evaluate the loss function only point-wise $L((\mathbf{x}^{(t)}, y^{(t)}), h)$ for the specific hypothesis h that has been used to compute the prediction $\hat{y}^{(t)} := h(\mathbf{x}^{(t)})$ at time instant t. This is fundamentally different from linear regression that uses the squared error loss (2.8) which can be evaluated for every possible hypothesis $h \in \mathcal{H}$.

3.15 LinUCB

ML methods are instrumental for various recommender systems [14]. A basic form of a recommender system amount to chose at some time instant t the most suitable item (product, song, movie) among a finite set of alternatives $a = 1, \ldots, A$. Each alternative is characterized by a feature vector $\mathbf{x}^{(t,a)}$ that varies between different time instants.

The data points arising in recommender systems typically represent different time instants t at which recommendations are computed. The data point at time t is characterized by a feature vector

$$\mathbf{x}^{(t)} = \left(\left(\mathbf{x}^{(t,1)}\right)^T, \ldots, \left(\mathbf{x}^{(t,A)}\right)^T \right)^T. \tag{3.27}$$

The feature vector $\mathbf{x}^{(t)}$ is obtained by stacking the feature vectors of alternatives at time t into a single long feature vector. The label of the data point t is a vector of rewards $\mathbf{y}^{(t)} := \left(r_1^{(t)}, \ldots, r_A^{(t)} \right)^T \in \mathbb{R}^A$. The entry $r_a^{(t)}$ represents the reward obtained by choosing (recommending) alternative a (with features $\mathbf{x}^{(t,a)}$) at time t. We might interpret the reward $r^{(t,a)}$ as an indicator if the customer actually buys the product corresponding to the recommended alternative a.

The ML method LinUCB (the name seems to be inspired by the terms "linear" and "upper confidence bound" (UCB)) aims at learning a hypothesis h that allows to predict the rewards $\mathbf{y}^{(i)}$ based on the feature vector $\mathbf{x}^{(t)}$ (3.27). As its hypothesis space $\mathcal{H}$, LinUCB uses the space of linear maps from the stacked feature vectors $\mathbb{R}^{nA}$ to the space of reward vectors $\mathbb{R}^A$. This hypothesis space can be parametrized by matrices $\mathbf{W} \in \mathbb{R}^{A \times nA}$. Thus, LinUCB learns a hypothesis that computes predicted rewards via

$$\widehat{\mathbf{y}}^{(t)} := \mathbf{W}\mathbf{x}^{(t)}. \tag{3.28}$$

The entries of $\widehat{\mathbf{y}}^{(t)} = \left(\hat{r}_1^{(t)}, \ldots, \hat{r}_A^{(t)} \right)$ are predictions of the individual rewards $r^{(t,a)}$. It seems natural to recommend at time t the alternative a whose predicted reward is maximum. However, it turns out that this approach is sub-optimal as it prevents the recommender system from learning the optimal predictor map $\mathbf{W}$.

Loosely speaking, LinUCB tries out (explores) each alternative $a \in \{1, \ldots, A\}$ sufficiently often to obtain a sufficient amount of training data for learning a good weight matrix $\mathbf{W}$. At time t, LinUCB chooses the alternative $a^{(t)}$ that maximizes the quantity

$$\hat{r}_a^{(t)} + R(t, a), \, a = 1, \ldots, A. \tag{3.29}$$

We can think of the component $R(t, a)$ as a form of confidence interval. It is constructed such that (3.29) upper bounds the actual reward $r_a^{(t)}$ with a prescribed level of confidence (or probability). The confidence term $R(t, a)$ depends on the feature vectors $\mathbf{x}^{(t', a)}$ of the alternative a at previous time instants $t' < t$. Thus, at each time instant t, LinUCB chooses the alternative a that results in the largest upper confidence bound (UCB) (3.29) on the reward (hence the "UCB" in LinUCB). We refer to the relevant literature on sequential learning (and decision making) for more details on the LinUCB [14].

3.16 Exercises

Exercise 3.1 Logistic loss and Accuracy Sect. 3.6 discussed logistic regression as a ML method that learns a linear hypothesis map by minimizing the logistic loss (3.15). The logistic loss has computationally pleasant properties as it is smooth and convex. However, in some applications we might be ultimately interested in the accuracy or (equivalently) the average 0/1 loss (2.9). Can we upper bound the average 0/1 loss using the average logistic loss incurred by a given hypothesis on a given training set?

Exercise 3.2 How Many Neurons? Consider a predictor map $h(x)$ which is piecewise linear and consisting of 1000 pieces. Assume we want to represent this map by an artificial neural network using neurons with one hidden layer of neurons with a ReLU activation functions. The output layer consists of a single neuron with linear activation function. How many neurons must the artificial neural network contain at least?

Exercise 3.3 Linear Classifiers Consider datapoints characterized by feature vectors $\mathbf{x} \in \mathbb{R}^n$ and binary labels $y \in \{-1, 1\}$. We are interested in finding a good linear classifier which is such that the feature vectors resulting in $h(\mathbf{x}) = 1$ is a half-space. Which of the methods discussed in this chapter aim at learning a linear classifier?

Exercise 3.4 Data Dependent Hypothesis space Consider a ML application involving data points that are characterized by features $\mathbf{x} \in \mathbb{R}^6$ and a numeric label $y \in \mathbb{R}$. We learn a hypothesis by minimizing the average loss incurred on a training set $\mathcal{D} = \{(\mathbf{x}^{(1)}, y^{(1)}), \ldots, (\mathbf{x}^{(m)}, y^{(m)})\}$. Which of the following ML methods uses a hypothesis space that depends on the dataset $\mathcal{D}$?

- logistic regression
- linear regression
- k-NN

Exercise 3.5 Triangle. Consider the artificial neural network in Fig. 3.12 using the deep net activation function (see Fig. 3.13). Show that there is a particular choice

Fig. 3.15 A hypothesis map
$h : \mathbb{R} \to \mathbb{R}$ with the shape of
a triangle

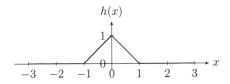

for the weights $\mathbf{w} = (w_1, \ldots, w_9)^T$ such that the resulting hypothesis map $h^{(\mathbf{w})}(x)$ is a triangle as depicted in Fig. 3.15. Can you also find a choice for the weights $\mathbf{w} = (w_1, \ldots, w_9)^T$ that produce the same triangle shape if we replace the deep net activation function with the linear function $f(z) = 10 \cdot z$?

Exercise 3.6 Approximate Triangles using Gaussians Try to approximate the hypothesis map depicted in Fig. 3.15 by an element of $\mathcal{H}_{\text{Gauss}}$ (see (3.14)) using $\sigma = 1/10, n = 10$ and $\mu_j = -1 + (2j/10)$.

Exercise 3.7 Privacy Leakage in k-NN Consider a k-NN method for a binary classification problem. We use $k = 1$ and a given training set whose data points characterize humans. Each human is characterized by a feature vector and label that indicates sensitive information (e.g., some sickness). Assume that you have access to the feature vectors of the training datapoints but not to the labels. Can you infer the label value of a training data point based on the prediction you are delivered based on your feature vector?

References

1. W. Rudin, *Principles of Mathematical Analysis*, 3rd edn. (McGraw-Hill, New York, 1976)
2. P.J. Huber, *Robust Statistics* (Wiley, New York, 1981)
3. M. Wainwright, *High-Dimensional Statistics: A Non-Asymptotic Viewpoint* (Cambridge University Press, Cambridge, 2019)
4. P. Bühlmann, S. van de Geer, *Statistics for High-Dimensional Data* (Springer, New York, 2011)
5. C. Lampert, Kernel methods in computer vision. Foundations and Trends in Computer Graphics and Vision **4**(3), 193–285 (2009)
6. T. Hastie, R. Tibshirani, J. Friedman, *The Elements of Statistical Learning* Springer Series in Statistics. (Springer, New York, 2001)
7. I. Goodfellow, Y. Bengio, A. Courville, *Deep Learning* (MIT Press, Cambridge, 2016)
8. G. Cybenko, Approximation by superpositions of a sigmoidal function. Math. Control Signals Systems **2**(4), 303–314 (1989)
9. R. Eldan and O. Shamir, The power of depth for feedforward neural networks. *CoRR*, abs/1512.03965 (2015)
10. E.L. Lehmann, G. Casella, *Theory of Point Estimation*, 2nd edn. (Springer, New York, 1998)

11. S. Levine, C. Finn, T. Darrell, P. Abbeel, End-to-end training of deep visuomotor policies. J. Mach. Learn. Res. **17**(1), 1334–1373 (2016)
12. R. Sutton, A. Barto, *Reinforcement Learning: An Introduction*, 2nd edn. (MIT Press, Cambridge, MA, 2018)
13. A. Ng, Shaping and Policy search in Reinforcement Learning. Ph.D. thesis, University of California, 2003
14. L. Li, W. Chu, J. Langford, and R. Schapire, A contextual-bandit approach to personalized news article recommendation, in *Proceedings of the International World Wide Web Conference*, pp. 661–670, 2010

Chapter 4
Empirical Risk Minimization

Chapter 2 discussed three main components of ML (see Fig. 2.1):

- data points characterized by features $\mathbf{x} \in \mathcal{X}$ and labels $y \in \mathcal{Y}$,
- a hypothesis space $\mathcal{H}$ of computationally feasible predictor maps $\mathcal{X} \to \mathcal{Y}$,
- and a loss function $L((\mathbf{x}, y), h)$ which measures the discrepancy between the predictions of a hypothesis h and actual data points

Ideally we would like to learn a hypothesis $h \in \mathcal{H}$ such that $L((\mathbf{x}, y), h)$ is small for any datapoint $(\mathbf{x}, y)$. However, in practice we can measure the loss only for a finite set of labeled datapoints, which serves as the training set. How can we know the loss of a hypothesis h when applied to datapoints outside the training set?

One possible approach to probe a hypothesis outside the training set is by using a **probabilistic model** for the data. Maybe the most widely used first choicie for such a probabilistic model is the i.i.d. assumption. Here, we interpret data points as realizations of i.i.d. RVs with a common probability distribution $p(\mathbf{x}, y)$. The training set is one particular set of such realizations drawn from $p(\mathbf{x}, y)$. Moreover, we can generate datapoints outside the training set by drawing realizations from the distribution $p(\mathbf{x}, y)$. Given this probability distribution over different realizations of datapoints allows us to define the risk of a hypothesis h as the expectation of the loss incurred by h on a random datapoint.

If we would know the probability distribution $p(\mathbf{x}, y)$, from which the datapoints are drawn, we could minimize the risk using probability theory. The optimal hypothesis, which is referred to as a Bayes estimator, can be read off directly from the posterior probability distribution $p(y|\mathbf{x})$ of the label y given the features $\mathbf{x}$ of a data point. The precise form of the Bayes estimator depends also on the choice for the lossfunc. When using the squared error loss, the optimal hypothesis (or Bayes estimator) is given by the posterior mean $h(\mathbf{x}) = \mathbb{E}\{y|\mathbf{x}\}$.

In most ML application, we do not know the true underlying probability distribution $p(\mathbf{x}, y)$ and have to estimate it from data. Therefore, we cannot compute the

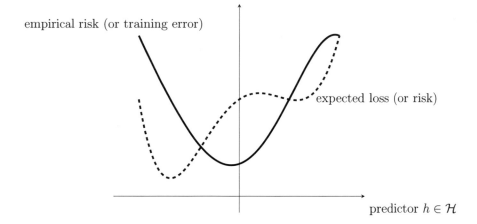

empirical risk (or training error)

expected loss (or risk)

predictor $h \in \mathcal{H}$

Fig. 4.1 ML methods learn a hypothesis $h \in \mathcal{H}$ that incur small loss when predicting the label y of datapoint based on its features $\mathbf{x}$. Empirical risk minimization approximates the expected loss or risk by the empirical risk (solid curve) incurred on a finite set of labeled datapoints (the training set). Note that we can compute the empirical risk based on the observed datapoints. However, to compute the risk we would need to know the underlying probability distribution which is rarely the case

Bayes estimator exactly. However, we can approximately compute this estimator by replacing the exact probability distribution with an estimate or approximation. Moreover, the risk of the Bayes estimator (which is the Bayes risk) provides a useful baseline against which we can compare the average loss incurred by a ML method on a set of data points. Sections 6.6 shows how to diagnose ML methods by comparing the average loss of a hypothesis on a training set and a validation set with a baseline.

Section 4.1 motivates empirical risk minimization by approximating the risk using the empirical risk (or average loss) computed for a set of labeled (training) datapoints (see Fig. 4.1). This approximation is justified by the law of large numbers which characterizes the deviation between averages of RVs and their expectation. Section 4.2 discusses the statistical and computational aspects of empirical risk minimization. We then specialize the empirical risk minimization for three particular ML methods arising from different combinations of hypothesis space and loss functions. Section 4.3 discusses empirical risk minimization for linear regression (see Sect. 3.1). Here, empirical risk minimization amounts to minimizing a differentiable convex function, which can be done efficiently using gradient-based methods (see Chap. 5).

We then discuss in Sect. 4.4 the empirical risk minimization obtained for decision tree models. The resulting empirical risk minimization problems becomes a discrete optimization problem which are typically much harder than convex optimization problems. We cannot apply gradient-based methods to solve the empirical risk minimization for decision trees. To solve the decision tree empirical risk minimization we essentially must try out all possible choices for the tree structure [1].

Section 4.5 considers the empirical risk minimization obtained when learning a linear hypothesis using the $0/1$ loss for classification problems. The resulting empirical risk minimization amounts to minimizing a non-differentiable and non-convex function. Instead of using computationally expensive methods for minimizing this function, we will use a different route via probability theory to construct approximate solutions to this empirical risk minimization instance.

As explained in Sect. 4.6, many ML methods use the empirical risk minimization during a training period to learn a hypothesis which is then applied to new datapoints during the inference period. Section 4.7 demonstrates how an online learning method can be obtained by solving the empirical risk minimization sequentially as new datapoints come in. Online learning methods continuously alternate between training and inference periods.

4.1 The Basic Idea of Empirical Risk Minimization

Consider some ML application that generates datapoints, each of which is characterized by a feature vector $\mathbf{x}$ and a label y. It can be useful to interpret data points as realizations of i.i.d. RVs with a common (joint) probability distribution $p(\mathbf{x}, y)$ for the features $\mathbf{x}$ and label y. The probability distribution $p(\mathbf{x}, y)$ allows to define the expected loss or risk of a hypothesis $h \in \mathcal{H}$ as

$$\mathbb{E}\{L((\mathbf{x}, y), h)\}. \tag{4.1}$$

It seems reasonable to aim at learning a hypothesis h such that its risk (4.1) is minimal,

$$h^* := \underset{h \in \mathcal{H}}{\operatorname{argmin}} \mathbb{E}\{L((\mathbf{x}, y), h)\}. \tag{4.2}$$

We refer to any hypothesis h^* that achieves the minimum risk (4.2) as a Bayes estimator [2]. Note that the Bayes estimator h^* depends on both, the probability distribution $p(\mathbf{x}, y)$ and the loss function. When using the squared error loss (2.8) in (4.2), the Bayes estimator h^* is given by the posterior mean of y given the features $\mathbf{x}$ (see [3, Ch. 7]).

Risk minimization (4.2) cannot be used for the design of ML methods whenever we do not know the probability distribution $p(\mathbf{x}, y)$. If we do not know the probability distribution $p(\mathbf{x}, y)$, which is the rule for many ML applications, we cannot evaluate the expectation in (4.1). One exception to this rule is if the datapoints are synthetically generated by drawing realizations from a given probability distribution $p(\mathbf{x}, y)$.

The idea of empirical risk minimization is to approximate the expectation in (4.2) with an average loss (the empirical risk) incurred on a given set of data points. As discussed in Sect. 2.3.4, this approximation is justified by the law of large numbers. We obtain empirical risk minimization by replacing the risk in the minimization problem (4.2) with the empirical risk (2.16),

$$\hat{h} = \underset{h \in \mathcal{H}}{\operatorname{argmin}} \widehat{L}(h|\mathcal{D})$$

$$\overset{(2.16)}{=} \underset{h \in \mathcal{H}}{\operatorname{argmin}} (1/m) \sum_{i=1}^{m} L((\mathbf{x}^{(i)}, y^{(i)}), h). \tag{4.3}$$

ML methods solve empirical risk minimization (4.3) to learn (finding) a good predictor $\hat{h} \in \mathcal{H}$ by "training" it on the dataset $\mathcal{D} = \{(\mathbf{x}^{(i)}, y^{(i)})\}_{i=1}^{m}$. This dataset is referred to as the training set and contains data point for which we know the label values (see Sect. 2.1.2). From a mathematical point of view, empirical risk minimization (4.3) is an optimization problem [4]. The optimization domain in (4.3) is the hypothesis space $\mathcal{H}$ of a ML method, the objective or cost function is the empirical risk (2.16).

It is important to remember that empirical risk minimization (4.3) is motivated by the law of large numbers. The law of large numbers, in turn, is only useful ("kicks in") if data points behave like realizations of i.i.d. RVs. This i.i.d. assumption is one of the most widely used working assumptions for the design and analysis of ML methods. However, there are many important application domains involving data points that clearly violate this i.i.d. assumption. One example for non-i.i.d. data are time series that consist of temporally ordered (consecutive) data points [5, 6]. Each data point in a time series might represent a particular time period. Another example for non-i.i.d. data arises in active learning where ML methods actively choose (or query) new data points [7]. For a third example of non-i.i.d. data, we refer to federated learning (FL) applications that involve collections (networks) of data generators with different statistical properties [8–12]. The details of ML methods for non-i.i.d. data are beyond the scope of this book.

4.2 Computational and Statistical Aspects of ERM

Solving the optimization problem (4.3) provides two things. First, the minimizer $\hat{h}$ is a predictor which performs optimal on the training set $\mathcal{D}$. Second, the corresponding objective value $\widehat{L}(\hat{h}|\mathcal{D})$ (the "training error") can be used to estimate for the risk or expected loss of $\hat{h}$. However, as we will discuss in Chap. 7, for some datasets $\mathcal{D}$, the training error $\widehat{L}(\hat{h}|\mathcal{D})$ obtained for $\mathcal{D}$ can be very different from the expected loss (risk) of $\hat{h}$ when applied to new datapoints which are not contained in $\mathcal{D}$. For any given hypothesis h, the i.i.d. assumption implies that the training error $\widehat{L}(h|\mathcal{D})$ is only a noisy approximation of the risk $\mathbb{E}\{L((\mathbf{x}, y), h)\}$. The empirical risk minimization solution $\hat{h}$ is the minimizer of this noisy approximation and therefore in general different from the Bayes estimator which minimizes the risk. In particular, even if the hypothesis $\hat{h}$ delivered by empirical risk minimization (4.3) has small training error $\widehat{L}(\hat{h}|\mathcal{D})$, it might have unacceptable large risk $\mathbb{E}\{L((\mathbf{x}, y), \hat{h})\}$.

Many important ML methods use hypotheses that are parametrized by weight vector $\mathbf{w}$. For each possible weight vector, we obtain a hypothesis $h^{(\mathbf{w})}(\mathbf{x})$. Such a parametrization is used in linear regression which learns a linear hypotheses

$h^{(\mathbf{w})}(\mathbf{x}) = \mathbf{w}^T \mathbf{x}$ with some weight vector $\mathbf{w}$. Another example for such a parametrization is obtained from artificial neural networks with the weights assigned to inputs of individual neurons (see Fig. 3.11).

For ML methods that use a parameterized hypothesis $h^{(\mathbf{w})}(\mathbf{x})$, we can reformulate the optimization problem (4.3) as an optimization of the weight vector,

$$\widehat{\mathbf{w}} = \operatorname*{argmin}_{\mathbf{w} \in \mathbb{R}^n} f(\mathbf{w}) \text{ with } f(\mathbf{w}) := (1/m) \sum_{i=1}^{m} L((\mathbf{x}^{(i)}, y^{(i)}), h^{(\mathbf{w})}). \qquad (4.4)$$

The objective function $f(\mathbf{w})$ in (4.4) is the empirical risk $\widehat{L}(h^{(\mathbf{w})}|\mathcal{D})$ incurred by the hypothesis $h^{(\mathbf{w})}$ when applied to the datapoints in the dataset $\mathcal{D}$. The optimization problems (4.4) and (4.3) are fully equivalent. Given the optimal weight vector $\widehat{\mathbf{w}}$ solving (4.4), the hypothesis $h^{(\widehat{\mathbf{w}})}$ solves (4.3).

We can interpret empirical risk minimization (4.3) as a form of learning by "trial and error". An instructor (or supervisor) provides some snapshots $\mathbf{z}^{(i)}$ which are characterized by features $\mathbf{x}^{(i)}$ and associated with known labels $y^{(i)}$. The learner then uses a hypothesis h to guess the labels $y^{(i)}$ only from the features $\mathbf{x}^{(i)}$ of all training data points. We then determine average loss or training error $\widehat{L}(h|\mathcal{D})$ that is incurred by the predictions $\hat{y}^{(i)} = h(\mathbf{x}^{(i)})$. If the error $\widehat{L}(h|\mathcal{D})$ is too large, we should try out another hypothesis map h' different from h with the hope of achieving a smaller training error $\widehat{L}(h'|\mathcal{D})$.

We highlight that the precise shape of the objective function $f(\mathbf{w})$ in (4.4) depends heavily on the parametrization of the predictor functions. The parametrization is the precise rule that assigns a hypothesis map $h^{(\mathbf{w})}$ to a given weight vector $\mathbf{w}$. The shape of $f(\mathbf{w})$ depends also on the choice for the loss function $L((\mathbf{x}^{(i)}, y^{(i)}), h)$. As depicted in Fig. 4.2, the different combinations of parametrized hypothesis space and loss functions can result in objective functions with fundamentally different properties such that their optimization is more or less difficult.

The objective function $f(\mathbf{w})$ for the empirical risk minimization obtained for linear regression (see Sect. 3.1) is differentiable and convex and can therefore be minimized using simple gradient-based methods (see Chap. 5). In contrast, the objective function $f(\mathbf{w})$ of ERM obtained for least absolute deviation regression and the support vector machine (see Sects. 3.3 and 3.7) is non-differentiable but still convex. The minimization of such functions is more challenging but still tractable as there exist efficient convex optimization methods which do not require differentiability of the objective function [13].

The objective function $f(\mathbf{w})$ obtained for artificial neural network are typically highly non-convex with many local minima. The optimization of non-convex objective function is in general more difficult than optimizing convex objective functions. However, it turns out that despite the non-convexity, iterative gradient-based methods can still be successfully applied to solve the resulting empirical risk minimization [14]. Even more challenging is the empirical risk minimization obtained for decision trees or Bayes estimator. These ML problems involve non-differentiable and non-convex objective functions.

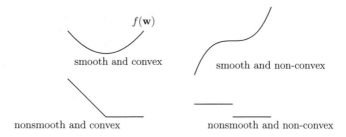

Fig. 4.2 Different types of objective functions that arise in empirical risk minimization for different combinations of hypothesis space and lossfunc

4.3 ERM for Linear Regression

As discussed in Sect. 3.1, linear regression methods learn a linear hypothesis $h^{(\mathbf{w})}(\mathbf{x}) = \mathbf{w}^T \mathbf{x}$ with minimum squared error loss (2.8). For linear regression, the empirical risk minimization problem (4.4) becomes

$$\widehat{\mathbf{w}} = \underset{\mathbf{w} \in \mathbb{R}^n}{\operatorname{argmin}} f(\mathbf{w})$$

$$\text{with } f(\mathbf{w}) := (1/m) \sum_{(\mathbf{x}, y) \in \mathcal{D}} (y - \mathbf{x}^T \mathbf{w})^2. \tag{4.5}$$

Here, $m = |\mathcal{D}|$ denotes the (sample-) size of the training set $\mathcal{D}$. The objective function $f(w)$ in (4.5) is computationally appealing since it is a convex and smooth function. Such a function can be minimized efficiently using the gradient-based methods discussed in Chap. 5.

We can rewrite the empirical risk minimization problem (4.5) more concisely by stacking the labels $y^{(i)}$ and feature vectors $\mathbf{x}^{(i)}$, for $i = 1, \ldots, m$, into a "label vector" $\mathbf{y}$ and "feature matrix" $\mathbf{X}$,

$$\mathbf{y} = (y^{(1)}, \ldots, y^{(m)})^T \in \mathbb{R}^m, \text{ and}$$

$$\mathbf{X} = \left(\mathbf{x}^{(1)}, \ldots, \mathbf{x}^{(m)}\right)^T \in \mathbb{R}^{m \times n}. \tag{4.6}$$

This allows us to rewrite the objective function in (4.5) as

$$f(\mathbf{w}) = (1/m) \|\mathbf{y} - \mathbf{X}\mathbf{w}\|_2^2. \tag{4.7}$$

Inserting (4.7) into (4.5), allows to rewrite the empirical risk minimization problem for linear regression as

$$\widehat{\mathbf{w}} = \underset{\mathbf{w} \in \mathbb{R}^n}{\operatorname{argmin}} (1/m) \|\mathbf{y} - \mathbf{X}\mathbf{w}\|_2^2. \tag{4.8}$$

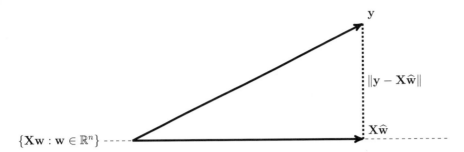

Fig. 4.3 The empirical risk minimization (4.8) for linear regression amounts to an orthogonal projection of the label vector $\mathbf{y} = \left(y^{(1)}, \ldots, y^{(m)}\right)^T$ on the subspace spanned by the columns of the feature matrix $\mathbf{X} = \left(\mathbf{x}^{(1)}, \ldots, \mathbf{x}^{(m)}\right)^T$

The formulation (4.8) allows for an interesting geometric interpretation of linear regression. Solving (4.8) amounts to finding a vector $\mathbf{Xw}$, with the feature matrix $\mathbf{X}$ (4.6), that is closest (in the Euclidean norm) to the label vector $\mathbf{y} \in \mathbb{R}^m$ (4.6). The solution to this approximation problem is precisely the orthogonal projection of the vector $\mathbf{y}$ onto the subspace of $\mathbb{R}^m$ that is spanned by the columns of the feature matrix $\mathbf{X}$ (see Fig. 4.3).

To solve the optimization problem (4.8), it is convenient to rewrite it as the quadratic problem

$$\min_{\mathbf{w} \in \mathbb{R}^n} \underbrace{(1/2)\mathbf{w}^T \mathbf{Qw} - \mathbf{q}^T \mathbf{w}}_{=f(\mathbf{w})}$$
$$\text{with } \mathbf{Q} = (1/m)\mathbf{X}^T \mathbf{X}, \mathbf{q} = (1/m)\mathbf{X}^T \mathbf{y}. \tag{4.9}$$

Since $f(\mathbf{w})$ is a differentiable and convex function, a necessary and sufficient condition for $\widehat{\mathbf{w}}$ to be a minimizer $f(\widehat{\mathbf{w}}) = \min_{\mathbf{w} \in \mathbb{R}^n} f(\mathbf{w})$ is the **zero-gradient condition** [4, Sec. 4.2.3]

$$\nabla f(\widehat{\mathbf{w}}) = \mathbf{0}. \tag{4.10}$$

Combining (4.9) with (4.10), yields the following necessary and sufficient condition for a weight vector $\widehat{\mathbf{w}}$ to solve the empirical risk minimization (4.5),

$$(1/m)\mathbf{X}^T \mathbf{X}\widehat{\mathbf{w}} = (1/m)\mathbf{X}^T \mathbf{y}. \tag{4.11}$$

This condition can be rewritten as

$$(1/m)\mathbf{X}^T \left(\mathbf{y} - \mathbf{X}\widehat{\mathbf{w}}\right) = \mathbf{0}. \tag{4.12}$$

We might refer to this condition as "normal equations" as they require the vector

$$\left(\mathbf{y} - \mathbf{X}\widehat{\mathbf{w}}\right) = \left(\left(y^{(1)} - \hat{y}^{(1)}\right), \ldots, \left(y^{(m)} - \hat{y}^{(m)}\right)\right)^T,$$

whose entries are the prediction errors for the datapoints in the training set, to be orthogonal (or normal) to the subspace spanned by the columns of the feature matrix $\mathbf{X}$.

It can be shown that, for any given feature matrix $\mathbf{X}$ and label vector $\mathbf{y}$, there always exists at least one optimal weight vector $\widehat{\mathbf{w}}$ which solves (4.11). The optimal weight vector might not be unique, i.e., there might be several different weight vectors achieving the minimum in (4.5). However, every vector $\widehat{\mathbf{w}}$ which solves (4.11) achieves the same minimum empirical risk

$$\widehat{L}(h^{(\widehat{\mathbf{w}})} \mid \mathcal{D}) = \min_{\mathbf{w} \in \mathbb{R}^n} \widehat{L}(h^{(\mathbf{w})} \mid \mathcal{D}) = \|(\mathbf{I} - \mathbf{P})\mathbf{y}\|^2. \tag{4.13}$$

Here, we used the orthogonal projection matrix $\mathbf{P} \in \mathbb{R}^{m \times m}$ on the linear span of the feature matrix $\mathbf{X} = (\mathbf{x}^{(1)}, \ldots, \mathbf{x}^{(m)})^T \in \mathbb{R}^{m \times n}$ (see (4.6)). The linear span of a matrix $\mathbf{A} = (\mathbf{a}^{(1)}, \ldots, \mathbf{a}^{(m)}) \in \mathbb{R}^{n \times m}$, denoted as span$\{\mathbf{A}\}$, is the subspace of $\mathbb{R}^n$ consisting of all linear combinations of the columns $\mathbf{a}^{(r)} \in \mathbb{R}^n$ of $\mathbf{A}$.

If the feature matrix $\mathbf{X}$ (see (4.6)) has full column rank, which implies that the matrix $\mathbf{X}^T\mathbf{X}$ is invertible, the projection matrix $\mathbf{P}$ is given explicitly as

$$\mathbf{P} = \mathbf{X}(\mathbf{X}^T\mathbf{X})^{-1}\mathbf{X}^T.$$

Moreover, the solution of (4.11) is then unique and given by

$$\widehat{\mathbf{w}} = (\mathbf{X}^T\mathbf{X})^{-1}\mathbf{X}^T\mathbf{y}. \tag{4.14}$$

The closed-form solution (4.14) requires the inversion of the $n \times n$ matrix $\mathbf{X}^T\mathbf{X}$.

Note that formula (4.14) is only valid if the matrix $\mathbf{X}^T\mathbf{X}$ is invertible. The feature matrix $\mathbf{X}$ is determined by the data points obtained in a ML application. Its properties are therefore not under the control of a ML method and it might well happen that the matrix $\mathbf{X}^T\mathbf{X}$ is not invertible. As a point in case, the matrix $\mathbf{X}^T\mathbf{X}$ cannot be invertible for any dataset containing fewer data points than the number of features used to characterize data points (this is referred to as high-dimensional data). Moreover, the matrix $\mathbf{X}^T\mathbf{X}$ is not invertible if there two co-linear features $x_j, x_{j'}$ such that $x_j = \beta x_{j'}$ holds for any data point with some constant $\alpha \in \mathbb{R}$.

Let us now consider a dataset such that the feature matrix $\mathbf{X}$ is not full column-rank and, in turn, the matrix $\mathbf{X}^T\mathbf{X}$ is not invertible. In this case we cannot use (4.14) to compute the optimal weight vector since the inverse of $\mathbf{X}^T\mathbf{X}$ does not exist. Moreover, in this case, there are infinitely many weight vectors that solve (4.11), i.e., the corresponding linear hypothesis map incurs minimum average squared error loss on the training set. Section 7.3 explains the benefits of using weights with small Euclidean norm. The weight vector $\widehat{\mathbf{w}}$ solving the linear regression optimality condition (4.11) and having minimum Euclidean norm among all such vectors is given by

$$\widehat{\mathbf{w}} = \left(\mathbf{X}^T\mathbf{X}\right)^{\dagger}\mathbf{X}^T\mathbf{y}. \tag{4.15}$$

Here, $\left(\mathbf{X}^T\mathbf{X}\right)^{\dagger}$ denotes the pseudoinverse (or the Moore–Penrose inverse) of $\mathbf{X}^T\mathbf{X}$ (see [15, 16]).

Computing the (pseudo-)inverse of $\mathbf{X}^T\mathbf{X}$ can be computationally challenging for large number n of features. Figure 2.4 depicts a simple ML problem where the number of features is already in the millions. The computational complexity of inverting the matrix $\mathbf{X}^T\mathbf{X}$ depends crucially on its condition number. We refer to a matrix as ill-conditioned if its condition number is much larger than 1. In general, ML methods do not have any control on the condition number of the matrix $\mathbf{X}^T\mathbf{X}$. Indeed, this matrix is determined solely by the (features of the) data points fed into the ML method.

Section 5.4 will discuss a method for computing the optimal weight vector $\widehat{\mathbf{w}}$ which does not require any matrix inversion. This method, referred to as gradient descent constructs a sequence $\mathbf{w}^{(0)}, \mathbf{w}^{(1)}, \ldots$ of increasingly accurate approximations of $\widehat{\mathbf{w}}$. This iterative method has two major benefits compared to evaluating the formula (4.14) using direct matrix inversion, such as Gauss-Jordan elimination [15].

First, GD typically requires significantly fewer arithmetic operations compared to direct matrix inversion. This is crucial in modern ML applications involving large feature matrices. Second, GD does not break when the matrix $\mathbf{X}$ is not full rank and the formula (4.14) cannot be used any more.

4.4 ERM for Decision Trees

Consider empirical risk minimization (4.3) for a regression problem with label space $\mathcal{Y} = \mathbb{R}$ and feature space $\mathcal{X} = \mathbb{R}^n$ and the hypothesis space defined by decision trees (see Sect. 3.10). In stark contrast to empirical risk minimization for linear regression or logistic regression, empirical risk minimization for decision trees amounts to a **discrete optimization problem**. Consider the particular hypothesis space $\mathcal{H}$ depicted in Fig. 3.9. This hypothesis space contains a finite number of different hypothesis maps. Each individual hypothesis map corresponds to a particular decision tree.

For the small hypothesis space $\mathcal{H}$ in Fig. 3.9, empirical risk minimization is easy. Indeed, we just have to evaluate the empirical risk ("training error") $\widehat{L}(h)$ for each hypothesis in $\mathcal{H}$ and pick the one yielding the smallest empirical risk. However, when allowing for a very large (deep) decision tree, the computational complexity of exactly solving the empirical risk minimization becomes intractable [17]. A popular approach to learn a decision tree is to use greedy algorithms which try to expand (grow) a given decision tree by adding new branches to leaf nodes in order to reduce the average loss on the training set (see [18, Chap. 8] for more details).

The idea behind many decision tree learning methods is quite simple: try out expanding a decision tree by replacing a leaf node with a decision node (implementing another "test" on the feature vector) in order to reduce the overall empirical risk much as possible.

Consider the labeled dataset $\mathcal{D}$ depicted in Fig. 4.4 and a given decision tree for predicting the label y based on the features $\mathbf{x}$. We might first try a hypothesis obtained from the simple tree shown in the top of Fig. 4.4. This hypothesis does not allow to achieve a small average loss on the training set $\mathcal{D}$. Therefore, we might grow the tree by replacing a leaf node with a decision node. According to Fig. 4.4, to so obtained larger decision tree provides a hypothesis that is able to perfectly predict the labels of the training set (it achieves zero empirical risk).

One important aspect of methods that learn a decision tree by sequentially growing the tree is the question of when to stop growing. A natural stopping criterion might be obtained from the limitations in computational resources, i.e., we can only afford to use decision trees up to certain maximum depth. Besides the computational limitations, we also face statistical limitations for the maximum size of decision trees. ML methods that allow for very deep decision trees, which represent highly complicated maps, tend

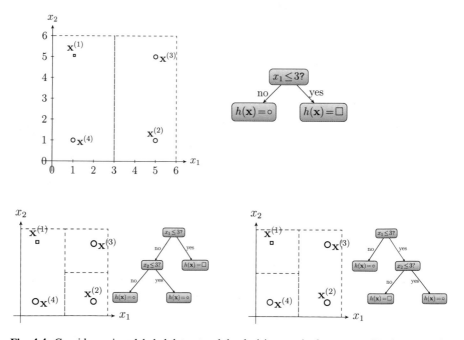

Fig. 4.4 Consider a given labeled dataset and the decision tree in the top row. We then grow the decision tree by expanding one of its two leaf nodes. The bottom row shows the resulting decision trees, along with their decision boundaries. Each decision tree in the bottom row is obtained by expanding a different leaf node of the decision tree in the top row

to overfit the training set (see Fig. 3.10 and Chap. 7). In particular, Even if a deep decision tree incurs small average loss on the training set, it might incur large loss when predicting the labels of data points outside the training set.

4.5 ERM for Bayes Classifiers

The family of ML methods referred to as Bayes estimator uses the $0/1$ loss (2.9) to measuring the quality of a classifier h. The resulting empirical risk minimization is

$$\hat{h} = \underset{h \in \mathcal{H}}{\operatorname{argmin}}(1/m) \sum_{i=1}^{m} L((\mathbf{x}^{(i)}, y^{(i)}), h)$$

$$\overset{(2.9)}{=} \underset{h \in \mathcal{H}}{\operatorname{argmin}}(1/m) \sum_{i=1}^{m} \mathcal{I}(h(\mathbf{x}^{(i)}) \neq y^{(i)}). \tag{4.16}$$

The objective function in this optimization problem is non-differentiable and non-convex (see Fig. 4.2). This prevents us from using gradient-based optimization methods (see Chap. 5) to solve (4.16).

We will now approach the empirical risk minimization (4.16) via a different route by interpreting the datapoints $(\mathbf{x}^{(i)}, y^{(i)})$ as realizations of i.i.d. RVs with the common probability distribution $p(\mathbf{x}, y)$.

As discussed in Sect. 2.3, the empirical risk obtained using $0/1$ loss approximates the error probability $p(\hat{y} \neq y)$ with the predicted label $\hat{y} = 1$ for $h(\mathbf{x}) > 0$ and $\hat{y} = -1$ otherwise (see (2.10)). Thus, we can approximate the empirical risk minimization (4.16) as

$$\hat{h} \overset{(2.10)}{\approx} \underset{h \in \mathcal{H}}{\operatorname{argmin}} \, p(\hat{y} \neq y). \tag{4.17}$$

Note that the hypothesis h, which is the optimization variable in (4.17), enters into the objective function of (4.17) via the definition of the predicted label $\hat{y}$, which is $\hat{y} = 1$ if $h(\mathbf{x}) > 0$ and $\hat{y} = -1$ otherwise.

It turns out that if we would know the probability distribution $p(\mathbf{x}, y)$, which is required to compute $p(\hat{y} \neq y)$, the solution of (4.17) can be found easily via elementary Bayesian decision theory [19]. In particular, the optimal classifier $h(\mathbf{x})$ is such that $\hat{y}$ achieves the maximum "a-posteriori" probability $p(\hat{y}|\mathbf{x})$ of the label being $\hat{y}$, given (or conditioned on) the features $\mathbf{x}$.

Since we typically do not know the probability distribution $p(\mathbf{x}, y)$, we have to estimate (or approximate) it from the observed data points $(\mathbf{x}^{(i)}, y^{(i)})$. This estimation is feasible if the data points can be considered (approximately) as realizations of i.i.d. RVs with a common joint distribution $p(\mathbf{x}, y)$. We can then estimate (the parameters) of the joint distribution $p(\mathbf{x}, y)$ using maximum likelihood methods (see Sect. 3.12).

For numeric features and labels, a widely-used parametric distribution $p(\mathbf{x}, y)$ is the multivariate normal (Gaussian) distribution. In particular, conditioned on the label y, the feature vector $\mathbf{x}$ is a Gaussian random vector with mean $\boldsymbol{\mu}_y$ and covariance Σ,[1]

$$p(\mathbf{x}|y) = \mathcal{N}(\mathbf{x}; \boldsymbol{\mu}_y, \Sigma). \tag{4.18}$$

The conditional expectation of the features $\mathbf{x}$, given (conditioned on) the label y of a data point, is $\boldsymbol{\mu}_1$ if $y = 1$, while for $y = -1$ the conditional mean of $\mathbf{x}$ is $\boldsymbol{\mu}_{-1}$. In contrast, the conditional covariance matrix $\Sigma = \mathbb{E}\{(\mathbf{x} - \boldsymbol{\mu}_y)(\mathbf{x} - \boldsymbol{\mu}_y)^T | y\}$ of $\mathbf{x}$ is the same for both values of the label $y \in \{-1, 1\}$. The conditional probability distribution $p(\mathbf{x}|y)$ of the feature vector, given the label y, is multivariate normal. In contrast, the marginal distribution of the features $\mathbf{x}$ is a Gaussian mixture model. We will revisit Gaussian mixture models later in Sect. 8.2 where we will see that they are a great tool for soft clustering.

For this probabilistic model of features and labels, the optimal classifier minimizing the error probability $p(\hat{y} \neq y)$ is $\hat{y} = 1$ for $h(\mathbf{x}) > 0$ and $\hat{y} = -1$ for $h(\mathbf{x}) \leq 0$ using the classifier map

$$h(\mathbf{x}) = \mathbf{w}^T \mathbf{x} \text{ with } \mathbf{w} = \Sigma^{-1}(\boldsymbol{\mu}_1 - \boldsymbol{\mu}_{-1}). \tag{4.19}$$

Carefully note that this expression is only valid if the matrix Σ is invertible.

We cannot implement the classifier (4.19) directly, since we do not know the true values of the class-specific mean vectors $\boldsymbol{\mu}_1, \boldsymbol{\mu}_{-1}$ and covariance matrix Σ. Therefore, we have to replace those unknown parameters with some estimates $\hat{\boldsymbol{\mu}}_1, \hat{\boldsymbol{\mu}}_{-1}$ and $\widehat{\Sigma}$. A principled approach is to use the maximum likelihood estimates (see (3.26))

$$\hat{\boldsymbol{\mu}}_1 = (1/m_1) \sum_{i=1}^{m} \mathcal{I}(y^{(i)} = 1)\mathbf{x}^{(i)},$$

$$\hat{\boldsymbol{\mu}}_{-1} = (1/m_{-1}) \sum_{i=1}^{m} \mathcal{I}(y^{(i)} = -1)\mathbf{x}^{(i)},$$

$$\hat{\boldsymbol{\mu}} = (1/m) \sum_{i=1}^{m} \mathbf{x}^{(i)},$$

$$\text{and } \widehat{\Sigma} = (1/m) \sum_{i=1}^{m} (\mathbf{z}^{(i)} - \hat{\boldsymbol{\mu}})(\mathbf{z}^{(i)} - \hat{\boldsymbol{\mu}})^T, \tag{4.20}$$

---

[1] We use the shorthand $\mathcal{N}(\mathbf{x}; \boldsymbol{\mu}, \Sigma)$ to denote the probability density function

$$p(\mathbf{x}) = \frac{1}{\sqrt{\det(2\pi\Sigma)}} \exp\left(- (1/2)(\mathbf{x}-\boldsymbol{\mu})^T \Sigma^{-1}(\mathbf{x}-\boldsymbol{\mu}) \right)$$

of a Gaussian random vector $\mathbf{x}$ with mean $\boldsymbol{\mu} = \mathbb{E}\{\mathbf{x}\}$ and covariance matrix $\Sigma = \mathbb{E}\{(\mathbf{x}-\boldsymbol{\mu})(\mathbf{x}-\boldsymbol{\mu})^T\}$.

with $m_1 = \sum_{i=1}^{m} \mathcal{I}(y^{(i)} = 1)$ denoting the number of datapoints with label $y = 1$ (m_{-1} is defined similarly). Inserting the estimates (4.20) into (4.19) yields the implementable classifier

$$h(\mathbf{x}) = \mathbf{w}^T \mathbf{x} \text{ with } \mathbf{w} = \widehat{\Sigma}^{-1}(\hat{\boldsymbol{\mu}}_1 - \hat{\boldsymbol{\mu}}_{-1}). \tag{4.21}$$

We highlight that the classifier (4.21) is only well-defined if the estimated covariance matrix $\widehat{\Sigma}$ (4.20) is invertible. This requires to use a sufficiently large number of training datapoints such that $m \geq n$.

We derived the classifier (4.21) as an approximate solution to the empirical risk minimization (4.16). The classifier (4.21) partitions the feature space $\mathbb{R}^n$ into two half-spaces. One half-space consists of feature vectors $\mathbf{x}$ for which the hypothesis (4.21) is non-negative and, in turn, $\hat{y} = 1$. The other half-space is constituted by feature vectors $\mathbf{x}$ for which the hypothesis (4.21) is negative and, in turn, $\hat{y} = -1$. Figure 2.9 illustrates these two half-spaces and the decision boundary between them.

The Bayes estimator (4.21) is another instance of a linear classifier like logistic regression and the support vector machine. Each of these methods learns a linear hypothesis $h(\mathbf{x}) = \mathbf{w}^T \mathbf{x}$, whose decision boundary (vectors $\mathbf{x}$ with $h(\mathbf{x}) = 0$) is a hyperplane (see Fig. 2.9). However, these methods use different loss functions for assessing the quality of a particular linear hypothesis $h(\mathbf{x}) = \mathbf{w}\mathbf{x}$ (which defined the decision boundary via $h(\mathbf{x}) = 0$). Therefore, these three methods typically learn classifiers with different decision boundaries.

For the estimator $\widehat{\Sigma}$ (3.26) to be accurate (close to the unknown covariance matrix) we need a number of datapoints (sample size) which is at least of the order n^2. This sample size requirement might be infeasible for applications with only few datapoints available.

The maximum likelihood estimate $\widehat{\Sigma}$ (4.20) is not invertible whenever $m < n$. In this case, the expression (4.21) becomes useless. To cope with small sample size $m < n$ we can simplify the model (4.18) by requiring the covariance to be diagonal $\Sigma = \text{diag}(\sigma_1^2, \ldots, \sigma_n^2)$. This is equivalent to modelling the individual features $x_1, \ldots, x_n$ of a datapoint as conditionally independent, given its label y. The resulting special case of a Bayes estimator is often referred to as a "naive Bayes" classifier.

We finally highlight that the classifier (4.21) is obtained using the generative model (4.18) for the data. Therefore, Bayes estimator belong to the family of generative ML methods which involve modelling the data generation. In contrast, logistic regression and the support vector machine do not require a generative model for the datapoints but aim directly at finding the relation between features $\mathbf{x}$ and label y of a data point. These methods belong therefore to the family of discriminative ML methods.

Generative methods such as those learning a Bayes estimator are preferable for applications with only very limited amounts of labeled data. Indeed, having a generative model such as (4.18) allows us to synthetically generate more labeled data by generating random features and labels according to the probability distribution (4.18). We refer to [20] for a more detailed comparison between generative and discriminative methods.

4.6   Training and Inference Periods

Some ML methods repeat the cycle in Figure 1 in a highly irregular fashion. Consider a large image collection which we use to learn a hypothesis about how cat images look like. It might be reasonable to adjust the hypothesis by fitting a model to the image collection. This fitting or training amounts to repeating the cycle in Figure 1 during some specific time period (the "training time") for a large number.

After the training period, we only apply the hypothesis to predict the labels of new images. This second phase is also known as inference time and might be much longer compared to the training time. Ideally, we would like to only have a very short training period to learn a good hypothesis and then only use the hypothesis for inference.

4.7   Online Learning

In it most basic form, empirical risk minimization requires a given set of labeled data points, which we refer to as the training set. However, some ML methods can access data only in a sequential fashion. As a point in case, consider time series data such as daily minimum and maximum temperatures recorded by a Finnish Meteorological Institute weather station. Such a time series consists of a sequence of data points that are generated at successive time instants.

Online learning studies ML methods that learn (or optimize) a hypothesis incrementally as new data arrives. This mode of operation is quite different from ML methods that learn a hypothesis at once by solving an empirical risk minimization problem. These different operation modes corresponds to different frequencies of iterating the basic ML cycle depicted in Figure 1. Online learning methods start a new cycle in Figure 1 whenever a new data point arrives (e.g., we have recorded the minimum and maximum temperate of a day that just ended).

We now present an online learning variant of linear regression (see Sect. 3.1) which is suitable for time series data with data points $\left(\mathbf{x}^{(t)}, y^{(t)}\right)$ gathered sequentially (over time). In particular, the data points $\left(\mathbf{x}^{(t)}, y^{(t)}\right)$ become available (are gathered) at a discrete time instants $t = 1, 2, 3 \ldots$.

Let us stack the feature vectors and labels of all data points available at time t into feature matrix $\mathbf{X}^{(t)}$ and label vector $\mathbf{y}^{(t)}$, respectively. The feature matrix and label vector for the first three time instants are

$$t = 1: \qquad \mathbf{X}^{(1)} := \left(\mathbf{x}^{(1)}\right)^{T}, \qquad\qquad \mathbf{y}^{(1)} = \left(y^{(1)}\right)^{T}, \qquad\qquad (4.22)$$

$$t = 2: \qquad \mathbf{X}^{(2)} := \left(\mathbf{x}^{(1)}, \mathbf{x}^{(2)}\right)^{T}, \qquad \mathbf{y}^{(2)} = \left(y^{(1)}, y^{(2)}\right)^{T}, \qquad (4.23)$$

$$t = 3: \qquad \mathbf{X}^{(3)} := \left(\mathbf{x}^{(1)}, \mathbf{x}^{(2)}, \mathbf{x}^{(3)}\right)^{T}, \qquad \mathbf{y}^{(3} = \left(y^{(1)}, y^{(2)}, y^{(3)}\right)^{T}. \qquad (4.24)$$

As detailed in Sect. 3.1, linear regression aims at learning the weights $\mathbf{w}$ of a linear map $h(\mathbf{x}) := \mathbf{w}^T \mathbf{x}$ such that the squared error loss $(y - h(\mathbf{x}))$ is as small as possible. This informal goal of linear regression is made precise by the empirical risk minimization problem (4.5) which defines the optimal weights via incurring minimum average squared error loss (empirical risk) on a given training set $\mathcal{D}$. These optimal weights are given by the solutions of (4.12). When the feature vectors of datapoints in $\mathcal{D}$ are linearly independent, we obtain the closed-form expression (4.14) for the optimal weights.

Inserting the feature matrix $\mathbf{X}^{(t)}$ and label vector $\mathbf{y}^{(t)}$ (4.22) into (4.14), yields

$$\widehat{\mathbf{w}}^{(t)} = \left(\left(\mathbf{X}^{(t)}\right)^T \mathbf{X}^{(t)}\right)^{-1} \left(\mathbf{X}^{(t)}\right)^T \mathbf{y}^{(t)}. \tag{4.25}$$

For each time instant we can evaluate the RHS of (4.25) to obtain the weight vector $\widehat{\mathbf{w}}^{(t)}$ that minimizes the average squared error loss over all data points gathered up to time t. However, computing $\widehat{\mathbf{w}}^{(t)}$ via direct evaluation of the RHS in (4.25) for each new time instant t misses an opportunity for recycling computations done already at earlier time instants.

Let us now show how to (partially) reuse the computations used to evaluate (4.25) for time t in the evaluation of (4.25) for the next time instant $t + 1$. To this end, we first rewrite the matrix $\mathbf{Q}^{(t)} := \left(\mathbf{X}^{(t)}\right)^T \mathbf{X}^{(t)}$ as

$$\mathbf{Q}^{(t)} = \sum_{r=1}^{t} \mathbf{x}^{(r)} \left(\mathbf{x}^{(r)}\right)^T. \tag{4.26}$$

Since $\mathbf{Q}^{(t+1)} = \mathbf{Q}^{(t)} + \mathbf{x}^{(t+1)} \left(\mathbf{x}^{(t+1)}\right)^T$, we can use a well-known identity for matrix inverses (see [21, 22]) to obtain

$$\left(\mathbf{Q}^{(t+1)}\right)^{-1} = \left(\mathbf{Q}^{(t)}\right)^{-1} + \frac{\left(\mathbf{Q}^{(t)}\right)^{-1} \mathbf{x}^{(t+1)} \left(\mathbf{x}^{(t+1)}\right)^T \left(\mathbf{Q}^{(t)}\right)^{-1}}{1 - \left(\mathbf{x}^{(t+1)}\right)^T \left(\mathbf{Q}^{(t)}\right)^{-1} \mathbf{x}^{(t+1)}}. \tag{4.27}$$

Inserting (4.27) into (4.25) yields the following relation between optimal weight vectors at consecutive time instants t and $t + 1$,

$$\widehat{\mathbf{w}}^{(t+1)} = \widehat{\mathbf{w}}^{(t)} - \left(\mathbf{Q}^{(t+1)}\right)^{-1} \mathbf{x}^{(t+1)} \left(\left(\mathbf{x}^{(t+1)}\right)^T \widehat{\mathbf{w}}^{(t)} - y^{(t+1)}\right). \tag{4.28}$$

Note that neither evaluating the RHS of (4.28) nor evaluating the RHS of (4.27) requires to actually invert a matrix of with more than one entry (we can think of a scalar number as 1×1 matrix). In contrast, evaluating the RHS (4.25) requires to invert the matrix $\mathbf{Q}^{(t)} \in \mathbb{R}^{n \times n}$. We obtain an online algorithm for linear regression via computing the updates (4.28) and (4.27) for each new time instant t. Another online method for linear regression will be discussed at the end of Sect. 5.7.

4.8 Exercise

Exercise 4.1 (Uniqueness in Linear Regression) What conditions on a training set ensure that there is a unique optimal linear hypothesis maps for linear regression.

Exercise 4.2 (Uniqueness in Linear Regression II) Linear regression uses the squared error loss (2.8) to measure the quality of a linear hypothesis map. We learn the weights $\mathbf{w}$ of a linear map via empirical risk minimization using a training set $\mathcal{D}$ that consists of $m = 100$ data points. Each data point is characterized by $n = 5$ features and a numeric label. Is there a unique choice for the weights $\mathbf{w}$ that results in a linear predictor with minimum average squared error loss on the training set $\mathcal{D})$?

Exercise 4.3 (A Simple Linear Regression Problem.) Consider a training set of m datapoints, each characterized by a single numeric feature x and numeric label y. We learn hypothesis map of the form $h(x) = x + b$ with some bias $b \in \mathbb{R}$. Can you write down a formula for the optimal b, that minimizes the average squared error on training data $\left(x^{(1)}, y^{(1)}\right), \ldots, \left(x^{(m)}, y^{(m)}\right)$.

Exercise 4.4 (Simple Least Absolute Deviation Problem.) Consider datapoints characterized by single numeric feature x and label y. We learn a hypothesis map of the form $h(x) = x + b$ with some bias $b \in \mathbb{R}$. Can you write down a formula for the optimal b, that minimizes the average absolute error on training data $\left(x^{(1)}, y^{(1)}\right), \ldots, \left(x^{(m)}, y^{(m)}\right)$.

Exercise 4.5 (Polynomial Regression.) Consider polynomial regression for datapoints with a single numeric feature $x \in \mathbb{R}$ and numeric label y. Here, polynomial regression is equivalent to linear regression using the transformed feature vectors $\mathbf{x} = \left(x^0, x^1, \ldots, x^{n-1}\right)^T$. Given a dataset $\mathcal{D} = \left(x^{(1)}, y^{(1)}\right), \ldots, \left(x^{(m)}, y^{(m)}\right)$, we construct the feature matrix $\mathbf{X} = \left(\mathbf{x}^{(1)}, \ldots, \mathbf{x}^{(m)}\right) \in \mathbb{R}^{m \times m}$ with its ith column given by the feature vector $\mathbf{x}^{(i)}$. Verify that this feature matrix is a Vandermonde matrix [23]? How is the determinant of the feature matrix related to the features and labels of data points in the dataset $\mathcal{D}$?

Exercise 4.6 (Training Error is not Expected Loss.) Consider a training set that consists of data points $\left(x^{(i)}, y^{(i)}\right)$, for $i = 1, \ldots, 100$, that are obtained as realizations of i.i.d. RVs. The common probability distribution of these RVs is defined by a random datapoint (x, y). The feature x of this random datapoint is a standard Gaussian RV with zero mean and unit variance. The label of a data point is modelled as $y = x + e$ with Gaussian noise $e \sim \mathcal{N}(0, 1)$. The feature x and noise e are statistically independent. We evaluate the specific hypothesis $h(x) = 0$ (which output 0 no matter what the feature value x is) by the training error $E_t = (1/m) \sum_{i=1}^{m} \left(y^{(i)} - h\left(x^{(i)}\right)\right)^2$. (which is the average squared error loss (2.8)). What is the probability that the training error E_t is at least 20 % larger than the expected (squared error) loss $\mathbb{E}\left\{(y - h(x))^2\right\}$? What is the mean (expected value) and variance of the training error ?

Exercise 4.7 (Optimization Methods as Filters.) Let us consider a fictional (idel) optimization method that can be represented as a filter $\mathcal{F}$. This filter $\mathcal{F}$ reads in a real-valued objective function $f(\cdot)$, defined for all weight vectors $\mathbf{w} \in \mathbb{R}^n$. The output of the filter $\mathcal{F}$ is another real-valued function $\hat{f}(\mathbf{w})$ that is defined point-wise as

$$\hat{f}(\mathbf{w}) = \begin{cases} 1, & \text{if } \mathbf{w} \text{ is a local minimum of } f(\cdot) \\ 0, & \text{otherwise.} \end{cases} \tag{4.29}$$

Verify that the filter $\mathcal{F}$ is shift or translation invariant, i.e., $\mathcal{F}$ commutes with a translation $f'(\mathbf{w}) := f(\mathbf{w} + \mathbf{w}^{(o)})$ with an arbitrary but fixed (reference) vector $\mathbf{w}^{(o)} \in \mathbb{R}^n$.

References

1. L. Hyafil, R. Rivest, Constructing optimal binary decision trees is np-complete. Inf. Process. Lett. **5**(1), 15–17 (1976)
2. E.L. Lehmann, G. Casella, *Theory of Point Estimation*, 2nd edn. (Springer, New York, 1998)
3. A. Papoulis, S.U. Pillai, *Probability, Random Variables, and Stochastic Processes*, 4th edn. (Mc-Graw Hill, New York, 2002)
4. S. Boyd, L. Vandenberghe, *Convex Optimization* (Cambridge University Press, Cambridge, UK, 2004)
5. P.J. Brockwell, R.A. Davis, *Time Series: Theory and Methods* (Springer, New York, 1991)
6. H. Lütkepohl, *New Introduction to Multiple Time Series Analysis* (Springer, New York, 2005)
7. D. Cohn, Z. Ghahramani, M. Jordan, Active learning with statistical models. J. Artif. Int. Res. **4**(1), 129–145 (1996). (March)
8. B. McMahan, E. Moore, D. Ramage, S. Hampson, B. A. y Arcas, Communication-efficient learning of deep networks from decentralized data, in *Proceedings of the 20th International Conference on Artificial Intelligence and Statistics*, volume 54 of *Proceedings of Machine Learning Research*, ed. by A. Singh and J. Zhu, pp. 1273–1282 (PMLR, 2017)
9. A. Jung, Networked exponential families for big data over networks. IEEE Access **8**, 202897–202909 (2020)
10. A. Jung, N. Tran, Localized linear regression in networked data. IEEE Sig. Proc. Lett. **26**(7), 1090–1094 (2019)
11. N. Tran, H. Ambos, A. Jung, Classifying partially labeled networked data via logistic network lasso, in *Proceedings of the IEEE International Conference on Acoustics, Speech and Signal Processing (ICASSP)*, pp. 3832–3836 (2020)
12. F. Sattler, K. Müller, and W. Samek. Clustered federated learning: Model-agnostic distributed multitask optimization under privacy constraints. *IEEE Transactions on Neural Networks and Learning Systems* (IEEE, New York, 2020)
13. N. Parikh, S. Boyd, Proximal algorithms. Foundations and Trends in Optimization **1**(3), 123–231 (2013)
14. I. Goodfellow, Y. Bengio, A. Courville, *Deep Learning* (MIT Press, Cambridge, 2016)
15. G.H. Golub, C.F. Van Loan, *Matrix Computations*, 3rd edn. (Johns Hopkins University Press, Baltimore, MD, 1996)
16. G. Golub, C. van Loan, An analysis of the total least squares problem. SIAM J. Numerical Analysis **17**(6), 883–893 (1980). (Dec.)
17. L. Hyafil, R.L. Rivest, Constructing optimal binary decision trees is np-complete. Inf. Process. Lett. **5**(1), 15–17 (1976)

18. G. James, D. Witten, T. Hastie, R. Tibshirani, *An Introduction to Statistical Learning with Applications in R* (Springer, Berlin, 2013)
19. H. Poor, *An Introduction to Signal Detection and Estimation*, 2nd edn. (Springer, Berlin, 1994)
20. A.Y. Ng, M.I. Jordan, On discriminative vs. generative classifiers: A comparison of logistic regression and naive bayes, in *Advances in Neural Information Processing Systems 14*. ed. by T.G. Dietterich, S. Becker, Z. Ghahramani (MIT Press, Cambridge, 2002), pp. 841–848
21. M.S. Bartlett, An inverse matrix adjustment arising in discriminant analysis. Ann. Math. Stat. **22**(1), 107–111 (1951)
22. C. Meyer, Generalized inversion of modified matrices. SIAM J. Appied Mathmetmatics **24**(3), 315–323 (1973)
23. W. Gautschi, G. Inglese, Lower bounds for the condition number of van der Monde matrices. Numer. Math. **52**, 241–250 (1988)

Chapter 5
Gradient-Based Learning

In what follows, we consider ML methods that use a parametrized hypothesis space $\mathcal{H}$. Each hypothesis $h^{(\mathbf{w})} \in \mathcal{H}$ in this space is characterized by a specific weight vector $\mathbf{w} \in \mathbb{R}^n$. Moreover, we consider ML methods that use a lossfunc $L((\mathbf{x}, y), h^{(\mathbf{w})})$ such that the average loss or empirical risk

$$f(\mathbf{w}) := (1/m) \sum_{i=1}^{m} L((\mathbf{x}^{(i)}, y^{(i)}), h^{(\mathbf{w})}) \tag{5.1}$$

depends smoothly on the weight vector $\mathbf{w}$.[1] This setting includes linear regression (see Sect. 3.1) and logistic regression (see Sect. 3.6).

The basic idea of ML methods is to learn a hypothesis whose predictions incur minimum loss. Section 4.1 made this idea precise using the principle of empirical risk minimization (4.4). empirical risk minimization (4.4) is an optimization problem that combines the three main components of ML. Indeed, the empirical risk minimization (4.4) binds together the data points in a training set $\mathcal{D}$, the hypothesis space $\mathcal{H}$ and the lossfunc $L((\mathbf{x}, y), h)$. To obtain practical ML methods we need to be able to efficiently solve empirical risk minimization (4.4) with a finite amount computational resources. These computational resources include storage capacity, communication bandwidth (for distributed or cloud computing) and processing time (which might is limited in real-time applications).

This chapter discusses gradient-based methods to solve empirical risk minimization (4.4). These are iterative methods that construct a sequence of weight vectors $\mathbf{w}^{(1)}, \ldots, \mathbf{w}^{(r)}$ such that the corresponding objective values $f(\mathbf{w}^{(1)}), \ldots, f(\mathbf{w}^{(r)})$ converge to the minimum $\min_{\mathbf{w} \in \mathbb{R}^n} f(\mathbf{w})$. The updates $\mathbf{w}^{(r)} \rightarrow \mathbf{w}^{(r+1)}$ between

---

[1] A function $f : \mathbb{R}^n \rightarrow \mathbb{R}$ is called smooth if it has continuous partial derivatives of all orders. In particular, we can define the gradient $\nabla f(\mathbf{w})$ for a smooth function $f(\mathbf{w})$ at every point $\mathbf{w}$.

© The Author(s), under exclusive license to Springer Nature Singapore Pte Ltd. 2022   99
A. Jung, *Machine Learning*, Machine Learning: Foundations, Methodologies, and Applications, https://doi.org/10.1007/978-981-16-8193-6_5

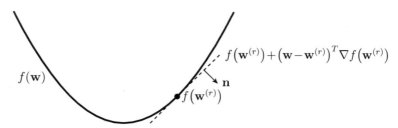

Fig. 5.1 A smooth function $f(\mathbf{w})$ can be approximated locally around a point $\mathbf{w}^{(r)}$ using a hyperplane whose normal vector $\mathbf{n} = (\nabla f(\mathbf{w}^{(r)}), -1)$ is determined by the gradient $\nabla f(\mathbf{w}^{(r)})$

consecutive weight vectors are based on so called GD steps. Section 5.1 discusses how the GD steps follows naturally from local linear approximations of the function $f(\mathbf{w})$ at the current iterate $\mathbf{w}^{(r)}$. These local linear approximations are constructed using the gradient of the function $f(\mathbf{w})$ at the current iterate $\mathbf{w}^{(r)}$. Gradient-based methods have gained popularity recently as an efficient technique for tuning the weights of deep nets within deep learning methods [1].

5.1   The GD Step

Consider a ML method that uses a parametrized hypothesis space and a smooth lossfunc such that the resulting empirical risk minimization (4.4) becomes a smooth optimization problem

$$\min_{\mathbf{w} \in \mathbb{R}^n} f(\mathbf{w}). \tag{5.2}$$

The smooth objective function $f : \mathbb{R}^n \to \mathbb{R}$ is the empirical risk (5.1) incurred by a hypothesis with weights $\mathbf{w} \in \mathbb{R}^n$. Our ultimate goal is to find a weight vector $\widehat{\mathbf{w}}$ that minimizes $f(\mathbf{w})$, $f(\widehat{\mathbf{w}}) = \min_{\mathbf{w} \in \mathbb{R}^n} f(\mathbf{w})$. However, for now we consider the simpler task of trying to improve a current guess or approximation $\mathbf{w}^{(r)}$ of an optimal weight vector $\widehat{\mathbf{w}}$. To this end, we approximate the objective function $f(\mathbf{w})$ by a simpler function. We will use this approximation only locally, in a sufficiently small neighbourhood of the current guess $\mathbf{w}^{(r)}$.

Since $f(\mathbf{w})$ is smooth, elementary calculus allows us to approximate it locally around some point $\mathbf{w}^{(r)}$ using a tangent hyperplane that passes through the point $(\mathbf{w}^{(r)}, f(\mathbf{w}^{(r)}))$. The normal vector of this hyperplance is given by $\mathbf{n} = (\nabla f(\mathbf{w}^{(r)}), -1)$ (see Fig. 5.1). The first component of the normal vector is the gradient of the function $f(\mathbf{w})$ at the point $\mathbf{w}^{(r)}$. Alternatively, we might define the gradient $\nabla f(\mathbf{w}^{(r)})$ via [2]

$$f(\mathbf{w}) \approx f(\mathbf{w}^{(r)}) + (\mathbf{w} - \mathbf{w}^{(r)})^T \nabla f(\mathbf{w}^{(r)}) \text{ for } \mathbf{w} \text{ sufficiently close to } \mathbf{w}^{(r)}. \tag{5.3}$$

Fig. 5.2 The GD step (5.4) updates the current weight vector $\mathbf{w}^{(r)}$ by adding the correction term $-\alpha\nabla f(\mathbf{w}^{(r)})$. The updated weight vector $\mathbf{w}^{(r+1)}$ is typically an improved approximation of the optimal weight vector

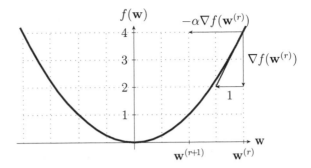

Remember that we would like to find a new (better) weight vector $\mathbf{w}^{(r+1)}$ that has smaller value $f(\mathbf{w}^{(r+1)}) < f(\mathbf{w}^{(r)})$ than the current guess $\mathbf{w}^{(r)}$. The approximation (5.3) suggests to choose the next guess $\mathbf{w} = \mathbf{w}^{(r+1)}$ such that $\big(\mathbf{w}^{(r+1)} - \mathbf{w}^{(r)}\big)^T \nabla f\big(\mathbf{w}^{(r)}\big)$ is negative. We can achieve this by the GD step

$$\mathbf{w}^{(r+1)} = \mathbf{w}^{(r)} - \alpha\nabla f(\mathbf{w}^{(r)}) \tag{5.4}$$

with a sufficiently small step size $\alpha > 0$. The step size α needs to be sufficiently small to ensure the validity linear approximation (5.3). In the context of ML, the GD step size parameter α is also referred to as learning rate. Indeed, the step size α determines the amount of progress during a GD step towards learning the optimal weight vector $\widehat{\mathbf{w}}$. However the interpretation of the step size as learning rate is only useful for sufficiently small step sizes. Indeed, when increasing the step size beyond a critical value, the iterates (5.4) depart from the optimal weight vector $\widehat{\mathbf{w}}$.

Gradient-based methods repeat the GD step (5.4) for a sufficient number of iterations (repetitions) to obtain an approximation to the optimal weight vector $\widehat{\mathbf{w}}$. For a convex differentiable objective function $f(\mathbf{w})$ and sufficiently small step size α, the iterates $f(\mathbf{w}^{(r)})$ obtained by repeating the gradient descent steps (5.4) converge to a minimum, i.e., $\lim_{r\to\infty} f(\mathbf{w}^{(r)}) = f(\widehat{\mathbf{w}})$ (see Fig. 5.2).

To implement the GD step (5.4) we need to choose a useful step size α. Moreover, executing the GD step (5.4) requires to compute the gradient $\nabla f(\mathbf{w}^{(r)})$. Both tasks can be challenging as discussed in Sects. 5.2 and 5.7. The success of deep learning methods, which represent predictor maps using artificial neural network (see Sec. 3.11), can be partially attributed to the ability of computing the gradient $\nabla f(\mathbf{w}^{(r)})$ efficiently via an algorithm known as **back-propagation** [1].

For the particular case of linear regression (see Sect. 5.4) and logistic regression (see Sect. 5.5), we will present precise conditions on the step size α which guarantee convergence of GD in Sects. 3.1 and 5.5. Moreover, for linear and logistic regression we can obtain closed-form expressions for the gradient $\nabla f(\mathbf{w})$ of the empirical risk (5.1). These closed-form expressions contain the feature vectors and labels of the data points in the training set $\mathcal{D} = \{(\mathbf{x}^{(1)}, y^{(1)}), \ldots, (\mathbf{x}^{(m)}, y^{(m)})\}$, which is used to calculate the empirical risk (5.1).

5.2   Choosing Step Size

The choice of the step size α in the GD step (5.4) has a significant impact on the performance of Algorithm 1. If we choose the step size α too large, the GD steps (5.4) diverge (see Fig. 5.3-(b)) and, in turn, Algorithm 1 fails to deliver a satisfactory approximation of the optimal weight vector $\mathbf{w}^{(\text{opt})}$ (see (5.7)).

If we choose the step size α too small (see Fig. 5.3-(a)), the updates (5.4) make only very little progress towards approximating the optimal weight vector $\widehat{\mathbf{w}}$. In applications that require real-time processing of data streams, it is possible to repeat the GD steps only for a moderate number. Thus If the step size is chosen too small, Algorithm 1 will fail to deliver a good approximation within an acceptable computation time.

The optimal choice of the step size α of GD can be a challenging task. Many sophisticated approaches for tuning the step size of gradient-based methods have been proposed [1, Chap. 8]. A discussion and analysis of these approaches is beyond the scope of this book. We will instead discuss sufficient condition on the step size which guarantee the convergence of the GD iterations to an optimum of (5.1).

Let us assume that the objective function $f(\mathbf{w})$ (5.1) is convex and smooth. Then, the GD steps (5.4) converge to an optimum of (5.1) for any step size α satisfying [3]

$$\alpha \leq \frac{1}{\lambda_{\max}\left(\nabla^2 f(\mathbf{w})\right)} \quad \text{for all } \mathbf{w} \in \mathbb{R}^n. \tag{5.5}$$

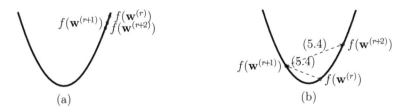

(a)   (b)

Fig. 5.3 Effect of choosing bad values for the learning rate α in the GD step(5.4). (a) If the steps size α in the GD step (5.4) is chosen too small, the iterations make very little progress towards the optimum or even fail to reach the optimum at all. (b) If the learning rate α is chosen too large, the iterates $\mathbf{w}^{(r)}$ might not converge at all (it might happen that $f(\mathbf{w}^{(r+1)}) > f(\mathbf{w}^{(r)})$!)

Here, we use the Hessian matrix $\nabla^2 f(\mathbf{w}) \in \mathbb{R}^{n \times n}$ of a smooth function $f(\mathbf{w})$ whose entries are the second-order partial derivatives $\frac{\partial f(\mathbf{w})}{\partial w_i \partial w_j}$ of the function $f(\mathbf{w})$. It is important to note that (5.5) guarantees convergence for every possible initialization $\mathbf{w}^{(0)}$ of the GD iterations.

Note that while it might be computationally challenging to determine the maximum (in absolute value) eigenvalue $\lambda_{\max}(\nabla^2 f(\mathbf{w}))$ for arbitrary $\mathbf{w}$, it might still be feasible to find an upper bound U for the maximum eigenvalue. If we know an upper bound $U \geq \lambda_{\max}(\nabla^2 f(\mathbf{w}))$ (valid for all $\mathbf{w} \in \mathbb{R}^n$), the step size $\alpha = 1/U$ still ensures convergence of the GD iteration.

5.3 When to Stop?

A main challenge in the successful application of GD is to decide when to stop iterating (or repeating) the GD step (5.4). One widely-used approach is to monitor the decrease in the objective function $f(\mathbf{w})$ and to stop if the decrease between the two most recent iterations falls below a threshold. Another stopping criterion is to use a fixed number of iterations (GD steps). However, for this to be effective we need some means to determine a sufficient number of iterations.

Let us assume that the objective function $f(\mathbf{w})$ is convex and we know upper and lower bounds for the eigenvalues of the Hessian $\nabla^2 f(\mathbf{w})$. It is then possible to derive upper bounds on the sub-optimality $f(\mathbf{w}^{(r)}) - \min_{\mathbf{w}} f(\mathbf{w})$ in terms of the number of GD steps used to obtain $\mathbf{w}^{(r)}$. These upper bounds allow then, in turn, to select the number of iterations such that the resulting sub-optimality is below a prescribed threshold.

5.4 GD for Linear Regression

Let us now use GD to implement some of the ML methods discussed in Chap. 3. In particular, we apply GD to linear regression as discussed in Section 3.1 to obtain a practical ML algorithm. This algorithm learns the weight vector for a linear hypothesis (see (3.1))

$$h^{(\mathbf{w})}(\mathbf{x}) = \mathbf{w}^T \mathbf{x}. \tag{5.6}$$

The weight vector is chosen to minimize average squared error loss (2.8)

$$\widehat{L}(h^{(\mathbf{w})}|\mathcal{D}) \overset{(4.4)}{=} (1/m) \sum_{i=1}^{m} (y^{(i)} - \mathbf{w}^T \mathbf{x}^{(i)})^2, \tag{5.7}$$

incurred by the predictor $h^{(\mathbf{w})}(\mathbf{x})$ when applied to the labeled dataset $\mathcal{D} = \{(\mathbf{x}^{(i)}, y^{(i)})\}_{i=1}^{m}$. The optimal weight vector $\widehat{\mathbf{w}}$ for (5.6) is characterized as

$$\widehat{\mathbf{w}} = \underset{\mathbf{w}\in\mathbb{R}^n}{\operatorname{argmin}} f(\mathbf{w}) \text{ with } f(\mathbf{w}) = (1/m) \sum_{i=1}^{m} \left(y^{(i)} - \mathbf{w}^T \mathbf{x}^{(i)}\right)^2. \qquad (5.8)$$

The optimization problem (5.8) is an instance of the smooth optimization problem (5.2). We can therefore use GD (5.4) to solve (5.8), to obtain the optimal weight vector $\widehat{\mathbf{w}}$. To implement GD, we need to compute the gradient $\nabla f(\mathbf{w})$.

The gradient of the objective function in (5.8) is given by

$$\nabla f(\mathbf{w}) = -(2/m) \sum_{i=1}^{m} \left(y^{(i)} - \mathbf{w}^T \mathbf{x}^{(i)}\right) \mathbf{x}^{(i)}. \qquad (5.9)$$

By inserting (5.9) into the basic GD iteration (5.4), we obtain Algorithm 1.

Algorithm 1 "Linear Regression via GD"

Input: labeled dataset $\mathcal{D} = \{(\mathbf{x}^{(i)}, y^{(i)})\}_{i=1}^{m}$ containing feature vectors $\mathbf{x}^{(i)} \in \mathbb{R}^n$ and labels $y^{(i)} \in \mathbb{R}$; GD learning rate $\alpha > 0$.
Initialize: set $\mathbf{w}^{(0)} := \mathbf{0}$; set iteration counter $r := 0$
1: **repeat**
2:    $r := r + 1$ (increase iteration counter)
3:    $\mathbf{w}^{(r)} := \mathbf{w}^{(r-1)} + \alpha(2/m) \sum_{i=1}^{m} \left(y^{(i)} - \left(\mathbf{w}^{(r-1)}\right)^T \mathbf{x}^{(i)}\right) \mathbf{x}^{(i)}$ (do a GD step (5.4))
4: **until** stopping criterion met
Output: $\mathbf{w}^{(r)}$ (which approximates $\widehat{\mathbf{w}}$ in (5.8))

Let us have a closer look on the update in step 3 of Algorithm 1, which is

$$\mathbf{w}^{(r)} := \mathbf{w}^{(r-1)} + \alpha(2/m) \sum_{i=1}^{m} \left(y^{(i)} - \left(\mathbf{w}^{(r-1)}\right)^T \mathbf{x}^{(i)}\right) \mathbf{x}^{(i)}. \qquad (5.10)$$

The update (5.10) has an appealing form as it amounts to correcting the previous guess (or approximation) $\mathbf{w}^{(r-1)}$ for the optimal weight vector $\widehat{\mathbf{w}}$ by the correction term

$$(2\alpha/m) \sum_{i=1}^{m} \underbrace{\left(y^{(i)} - \left(\mathbf{w}^{(r-1)}\right)^T \mathbf{x}^{(i)}\right)}_{e^{(i)}} \mathbf{x}^{(i)}. \qquad (5.11)$$

The correction term (5.11) is a weighted average of the feature vectors $\mathbf{x}^{(i)}$ using weights $(2\alpha/m) \cdot e^{(i)}$. These weights consist of the global factor $(2\alpha/m)$ (that applies equally to all feature vectors $\mathbf{x}^{(i)}$) and a sample-specific factor $e^{(i)} =$

$(y^{(i)} - (\mathbf{w}^{(r-1)})^T \mathbf{x}^{(i)})$, which is the prediction (approximation) error obtained by the linear predictor $h^{(\mathbf{w}^{(r-1)})}(\mathbf{x}^{(i)}) = (\mathbf{w}^{(r-1)})^T \mathbf{x}^{(i)}$ when predicting the label $y^{(i)}$ from the features $\mathbf{x}^{(i)}$.

We can interpret the GD step (5.10) as an instance of "learning by trial and error". Indeed, the GD step amounts to first "trying out" (trial) the predictor $h(\mathbf{x}^{(i)}) = (\mathbf{w}^{(r-1)})^T \mathbf{x}^{(i)}$. The predicted values are then used to correct the weight vector $\mathbf{w}^{(r-1)}$ according to the error $e^{(i)} = y^{(i)} - (\mathbf{w}^{(r-1)})^T \mathbf{x}^{(i)}$.

The choice of the learning rate α used for Algorithm 1 can be based on the condition (5.5) with the Hessian $\nabla^2 f(\mathbf{w})$ of the objective function $f(\mathbf{w})$ underlying linear regression (see (5.8)). This Hessian is given explicitly as

$$\nabla^2 f(\mathbf{w}) = (1/m)\mathbf{X}^T \mathbf{X}, \tag{5.12}$$

with the feature matrix $\mathbf{X} = (\mathbf{x}^{(1)}, \dots, \mathbf{x}^{(m)})^T \in \mathbb{R}^{m \times n}$ (see (4.6)). Note that the Hessian (5.12) does not depend on the weight vector $\mathbf{w}$.

Comparing (5.12) with (5.5), one particular strategy for choosing the step size in Algorithm 1 is to (i) compute the matrix product $\mathbf{X}^T \mathbf{X}$, (ii) compute the maximum eigenvalue $\lambda_{\max}((1/m)\mathbf{X}^T \mathbf{X})$ of this product and (iii) set the step size to $\alpha = 1/\lambda_{\max}((1/m)\mathbf{X}^T \mathbf{X})$.

While it might be challenging to compute the maximum eigenvalue $\lambda_{\max}((1/m)\mathbf{X}^T \mathbf{X})$, it might be easier to find an upper bound U for it.[2] Given such an upper bound $U \geq \lambda_{\max}((1/m)\mathbf{X}^T \mathbf{X})$, the step size $\alpha = 1/U$ still ensures convergence of the GD iteration. Consider a dataset $\{(\mathbf{x}^{(i)}, y^{(i)})\}_{i=1}^{m}$ with normalized features, i.e., $\|\mathbf{x}^{(i)}\| = 1$ for all $i = 1, \dots, m$. Then, by elementary linear algebra, one can verify the upper bound $U = 1$, i.e., $1 \geq \lambda_{\max}((1/m)\mathbf{X}^T \mathbf{X})$. We can then ensure convergence of the GD iterations $\mathbf{w}^{(r)}$ (see (5.10)) by choosing the step size $\alpha = 1$.

Time-Data Tradeoffs. The number of iteration required by Algorithm 1 to ensure a prescribed sub-optimality depends crucially on the condition number of $\mathbf{X}^T \mathbf{X}$. What can we say about the condition number? In general, we have not control over this quantity as the matrix $\mathbf{X}$ consists of the feature vectors of arbitrary datapoints. However, it is often useful to model the feature vectors as realizations of i.i.d. random vectors. It is then possible to bound the probability of the feature matrix having very small condition number. These bounds can then be used to choose the step-size such that convergence is guaranteed with sufficiently large probability. The usefulness of these bounds typically depends on the ratio n/m. For increasing sample-size, these bounds allow to use larger step-sizes and, in turn, result in faster convergence of GD. Thus, we obtain a trade-off between convergence time of GD and the number of data points. Such time-data trade-offs have been studied recently for linear regression with known structure of the weight vector [4]

---

[2] The problem of computing a full eigenvalue decomposition of $\mathbf{X}^T \mathbf{X}$ has essentially the same complexity as empirical risk minimization via directly solving (4.11), which we want to avoid by using the "cheaper" Algorithm 1.

5.5   GD for Logistic Regression

As discussed in Sect. 3.6, logistic regression learns a linear hypothesis $h^{(\widehat{\mathbf{w}})}$ by minimizing the average logistic loss (3.15) obtained for a dataset $\mathcal{D} = \{(\mathbf{x}^{(i)}, y^{(i)})\}_{i=1}^{m}$, with features $\mathbf{x}^{(i)} \in \mathbb{R}^n$ and binary labels $y^{(i)} \in \{-1, 1\}$. This minimization problem is an instance of the smooth optimization problem (5.2),

$$\widehat{\mathbf{w}} = \underset{\mathbf{w} \in \mathbb{R}^n}{\operatorname{argmin}} f(\mathbf{w})$$

$$\text{with } f(\mathbf{w}) = (1/m) \sum_{i=1}^{m} \log(1 + \exp(-y^{(i)} \mathbf{w}^T \mathbf{x}^{(i)})). \tag{5.13}$$

The application of the gradient step (5.4) to solve (5.13), requires computing the gradient $\nabla f(\mathbf{w})$. The gradient of the objective function in (5.13) is

$$\nabla f(\mathbf{w}) = (1/m) \sum_{i=1}^{m} \frac{-y^{(i)}}{1 + \exp(y^{(i)} \mathbf{w}^T \mathbf{x}^{(i)})} \mathbf{x}^{(i)}. \tag{5.14}$$

By inserting (5.14) into the basic GD iteration (5.4), we obtain Algorithm 2.

Algorithm 2 "Logistic regression via GD"

Input: labeled dataset $\mathcal{D} = \{(\mathbf{x}^{(i)}, y^{(i)})\}_{i=1}^{m}$ containing feature vectors $\mathbf{x}^{(i)} \in \mathbb{R}^n$ and labels $y^{(i)} \in \mathbb{R}$; GD learning rate $\alpha > 0$.
Initialize: set $\mathbf{w}^{(0)} := \mathbf{0}$; set iteration counter $r := 0$
1: **repeat**
2:     $r := r + 1$ (increase iteration counter)
3:     $\mathbf{w}^{(r)} := \mathbf{w}^{(r-1)} + \alpha(1/m) \sum_{i=1}^{m} \frac{y^{(i)}}{1 + \exp\left(y^{(i)} \left(\mathbf{w}^{(r-1)}\right)^T \mathbf{x}^{(i)}\right)} \mathbf{x}^{(i)}$ (do a gradient descent step (5.4))
4: **until** stopping criterion met
Output: $\mathbf{w}^{(r)}$, which approximates a solution $\widehat{\mathbf{w}}$ of (5.13))

Let us have a closer look on the update in step (3) of Algorithm 2. This step amounts to computing

$$\mathbf{w}^{(r)} := \mathbf{w}^{(r-1)} + \alpha(1/m) \sum_{i=1}^{m} \frac{y^{(i)}}{1 + \exp\left(y^{(i)} \left(\mathbf{w}^{(r-1)}\right)^T \mathbf{x}^{(i)}\right)} \mathbf{x}^{(i)}. \tag{5.15}$$

Similar to the GD step (5.10) for linear regression, also the gradient descent step (5.15) for logistic regression can be interpreted as an implementation of the trial-and-error principle. Indeed, (5.15) corrects the previous guess (or approximation) $\mathbf{w}^{(r-1)}$ for the optimal weight vector $\widehat{\mathbf{w}}$ by the correction term

$$(\alpha/m) \sum_{i=1}^{m} \underbrace{\frac{y^{(i)}}{1 + \exp(y^{(i)}\mathbf{w}^T\mathbf{x}^{(i)})}}_{e^{(i)}} \mathbf{x}^{(i)}. \tag{5.16}$$

The correction term (5.16) is a weighted average of the feature vectors $\mathbf{x}^{(i)}$. The feature vector $\mathbf{x}^{(i)}$ is weighted by the factor $(\alpha/m) \cdot e^{(i)}$. These weighting factors are a product of the global factor (α/m) that applies equally to all feature vectors $\mathbf{x}^{(i)}$. The global factor is multiplied by a datapoint-specific factor $e^{(i)} = \frac{y^{(i)}}{1+\exp(y^{(i)}\mathbf{w}^T\mathbf{x}^{(i)})}$, which quantifies the error of the classifier $h^{(\mathbf{w}^{(r-1)})}(\mathbf{x}^{(i)}) = (\mathbf{w}^{(r-1)})^T\mathbf{x}^{(i)}$ for a single data point with true label $y^{(i)} \in \{-1, 1\}$ and features $\mathbf{x}^{(i)} \in \mathbb{R}^n$.

We can use the sufficient condition (5.5) for the convergence of GD to guide the choice of the step size α in Algorithm 2. To apply condition (5.5), we need to determine the Hessian $\nabla^2 f(\mathbf{w})$ matrix of the objective function $f(\mathbf{w})$ underlying logistic regression (see (5.13)). Some basic calculus reveals (see [5, Ch. 4.4.])

$$\nabla^2 f(\mathbf{w}) = (1/m)\mathbf{X}^T\mathbf{D}\mathbf{X}. \tag{5.17}$$

Here, we used the feature matrix $\mathbf{X} = (\mathbf{x}^{(1)}, \ldots, \mathbf{x}^{(m)})^T \in \mathbb{R}^{m \times n}$ (see (4.6)) and the diagonal matrix $\mathbf{D} = \text{diag}\{d_1, \ldots, d_m\} \in \mathbb{R}^{m \times m}$ with diagonal elements

$$d_i = \frac{1}{1 + \exp(-\mathbf{w}^T\mathbf{x}^{(i)})}\left(1 - \frac{1}{1 + \exp(-\mathbf{w}^T\mathbf{x}^{(i)})}\right). \tag{5.18}$$

We highlight that, in contrast to the Hessian (5.12) of the objective function arising in linear regression, the Hessian (5.17) of logistic regression varies with the weight vector $\mathbf{w}$. This makes the analysis of Algorithm 2 and the optimal choice for the step size α more difficult compared to Algorithm 1. At least, we can ensure convergence of (5.15) (towards a solution of (5.13)) for the step size $\alpha = 1$ if we normalize feature vectors such that $\|\mathbf{x}^{(i)}\| = 1$. This follows from the fact the diagonal entries (5.18) take values in the interval $[0, 1]$.

5.6 Data Normalization

The convergence speed of the GD steps (5.4), i.e., the number of steps required to reach the minimum of the objective function (4.5) within a prescribed accuracy, depends crucially on the condition number $\kappa(\mathbf{X}^T\mathbf{X})$. This condition number is defined as the ratio

$$\kappa(\mathbf{X}^T\mathbf{X}) := \lambda_{\max}/\lambda_{\min} \tag{5.19}$$

between the largest and smallest eigenvalue of the matrix $\mathbf{X}^T\mathbf{X}$.

The condition number is only well-defined if the columns of the feature matrix $\mathbf{X}$ (see (4.6)), which are precisely the feature vectors $\mathbf{x}^{(i)}$, are linearly independent. In this case the condition number is lower bounded as $\kappa(\mathbf{X}^T\mathbf{X}) \geq 1$.

It can be shown that the GD steps (5.4) converge faster for smaller condition number $\kappa(\mathbf{X}^T\mathbf{X})$ [6]. Thus, GD will be faster for datasets with a feature matrix $\mathbf{X}$ such that $\kappa(\mathbf{X}^T\mathbf{X}) \approx 1$. It is therefore often beneficial to pre-process the feature vectors using a **normalization** (or **standardization**) procedure as detailed in Algorithm 3.

Algorithm 3 "Data Normalization"

Input: labeled dataset $\mathcal{D} = \{(\mathbf{x}^{(i)}, y^{(i)})\}_{i=1}^m$

1: remove sample means $\widehat{\mathbf{x}} = (1/m)\sum_{i=1}^m \mathbf{x}^{(i)}$ from features, i.e.,

$$\mathbf{x}^{(i)} := \mathbf{x}^{(i)} - \widehat{\mathbf{x}} \text{ for } i = 1, \ldots, m$$

2: normalise features to have unit variance,

$$\hat{x}_j^{(i)} := x_j^{(i)}/\hat{\sigma} \text{ for } j = 1, \ldots, n \text{ and } i = 1, \ldots, m$$

with the empirical (sample) variance $\hat{\sigma}_j^2 = (1/m)\sum_{i=1}^m \left(x_j^{(i)}\right)^2$

Output: normalized feature vectors $\{\widehat{\mathbf{x}}^{(i)}\}_{i=1}^m$

Algorithm 3 transforms the original feature vectors $\mathbf{x}^{(i)}$ into new feature vectors $\widehat{\mathbf{x}}^{(i)}$ such that the new feature matrix $\widehat{\mathbf{X}} = (\widehat{\mathbf{x}}^{(1)}, \ldots, \widehat{\mathbf{x}}^{(m)})^T$ is better conditioned than the original feature matrix, i.e., $\kappa(\widehat{\mathbf{X}}^T\widehat{\mathbf{X}}) < \kappa(\mathbf{X}^T\mathbf{X})$.

5.7 Stochastic GD

Consider the GD steps (5.4) for minimizing the empirical risk (5.1). The gradient $\nabla f(\mathbf{w})$ of the objective function (5.1) has a particular structure. Indeed, this gradient is a sum

$$\nabla f(\mathbf{w}) = (1/m)\sum_{i=1}^m \nabla f_i(\mathbf{w}) \text{ with } f_i(\mathbf{w}) := L((\mathbf{x}^{(i)}, y^{(i)}), h^{(\mathbf{w})}). \qquad (5.20)$$

Each component of the sum (5.20) corresponds to one particular data points $(\mathbf{x}^{(i)}, y^{(i)})$, for $i = 1, \ldots, m$. We need to compute a sum of the form (5.20) for each new GD step (5.4).

Computing the sum in (5.20) can be computationally challenging for at least two reasons. First, computing the sum is challenging for very large datasets with m in the order of billions. Second, for datasets which are stored in different data centres located all over the world, the summation would require a huge amount of network resources. Moreover, the finite transmission rate of communication networks limits the rate by which the GD steps (5.4) can be executed.

The idea of stochastic gradient descent (SGD) is to replace the exact gradient $\nabla f(\mathbf{w})$ (5.20) by an approximation that is easier to compute than a direct evaluation of (5.20). The word "stochastic" in the name stochastic GD hints already at the use of a stochastic approximation $g(\mathbf{w}) \approx \nabla f(\mathbf{w})$. It turns out that using a gradient approximation $g(\mathbf{w})$ can result in significant savings in computational complexity while sacrificing only a graceful degradation in the overall optimization accuracy. The optimization accuracy (distance to minimum of $f(\mathbf{w})$) depends crucially on the approximation error or "gradient noise"

$$\varepsilon := \nabla f(\mathbf{w}) - g(\mathbf{w}). \tag{5.21}$$

The elementary step of any stochastic GD method is obtained from the gradient descent step (5.4) by replacing the exact gradient $\nabla f(\mathbf{w})$ with its approximation $g(\mathbf{w})$,

$$\mathbf{w}^{(r+1)} = \mathbf{w}^{(r)} - \alpha_r g(\mathbf{w}^{(r)}), \tag{5.22}$$

As the notation in (5.22) indicates, stochastic gradient descent methods use a learning rate α_r that varies between different iterations.

To avoid a detrimental accumulation of the gradient noise (5.21) during the stochastic gradient descent updates (5.22), many stochastic gradient descent methods decrease the learning rate α as the iterations proceed. The sequence α_r of learning rate is referred to as a learning rate schedule [1, Chap. 8]. One possible choice for the learning rate schedule is $\alpha_r := 1/r$ [7]. Exercise 5.2 discusses conditions on the learning rate schedule that guarantee convergence of the updates stochastic gradient descent to the minimum of $f(w)$.

The approximate ("noisy") gradient $g(\mathbf{w})$ can be obtained by different randomization strategies. The most basic form of stochastic gradient descent constructs the gradient approximation g using a randomly selected component $\nabla f_i(\mathbf{w})$ in (5.20). The index $\hat{i}$ is chosen randomly from the set $\{1, \ldots, m\}$. The resulting Stochastic gradient descent method then repeats the update

$$\mathbf{w}^{(r+1)} = \mathbf{w}^{(r)} - \alpha \nabla f_{\hat{i}_r}(\mathbf{w}^{(r)}), \tag{5.23}$$

sufficiently often. Every update (5.23) uses a fresh randomly chosen index $\hat{i}_r$, i.e., the indices $\hat{i}_r$ used in different iterations are statistically independent.

Note that (5.23) replaces the summation over all training datapoints in the gradient descent step (5.4) just by the random selection of a single component of the sum. The resulting savings in computational complexity can be significant in applications where a large number of datapoints is stored in a distributed fashion. However, this saving in computational complexity comes at the cost of introducing a non-zero gradient noise

$$\varepsilon_r \overset{(5.21)}{=} \nabla f(\mathbf{w}^{(r)}) - g(\mathbf{w}^{(r)})$$
$$= \nabla f(\mathbf{w}^{(r)}) - \nabla f_{i_r}(\mathbf{w}). \tag{5.24}$$

Mini-Batch Stochastic gradient descent. Let us now discuss a variant of stochastic gradient descent that aims at reducing the approximation error (gradient noise) (5.24) occurring during the basic stochastic gradient descent update (5.23). The idea is to use more than one randomly selected component $\nabla f_i(\mathbf{w})$ (see (5.20)) for constructing a gradient approximation. In particular, given a batch size B, we randomly select a subset $\mathcal{B} = \{i_1, \ldots, i_B\}$ (a "batch") which is used to construct the gradient approximation

$$g(\mathbf{w}) = (1/B) \sum_{i' \in \mathcal{B}} \nabla f_{i'}(\mathbf{w}). \tag{5.25}$$

Algorithm 4 summarizes mini-batch stochastic gradient descent which uses the gradient approximation (5.25) in the generic stochastic gradient descent update (5.22).

Algorithm 4 Mini-Batch stochastic gradient descent

Input: components $f_i(\mathbf{w})$, for $i = 1, \ldots, m$ of objective function $f(\mathbf{w}) = \sum_{i=1}^{m} f_i(\mathbf{w})$; batch size B; learning rate schedule $\alpha_r > 0$.
Initialize: set $\mathbf{w}^{(0)} := \mathbf{0}$; set iteration counter $r := 0$
1: **repeat**
2:    randomly select a batch $\mathcal{B} = \{i_1, \ldots, i_B\} \subseteq \{1, \ldots, m\}$ of indices that select a subset of components f_i
3:    compute an approximate gradient $g(\mathbf{w}^{(r)})$ using (5.25)
4:    $r := r + 1$ (increase iteration counter)
5:    $\mathbf{w}^{(r)} := \mathbf{w}^{(r-1)} - \alpha_r g(\mathbf{w}^{(r-1)})$
6: **until** stopping criterion met
Output: $\mathbf{w}^{(r)}$ (which approximates $\text{argmin}_{\mathbf{w} \in \mathbb{R}^n} f(\mathbf{w})$))

Note that Algorithm 4 includes the basic stochastic gradient descent variant (5.23) as a special case when using batch size $B = 1$. Another special case of Algorithm 4 is gradient descent (5.4), which is obtained for the batch size $B = m$.

Online Learning with stochastic gradient descent. The basic stochastic gradient descent iteration (5.26) assumes that the training data is already collected but so large that the sum in (5.20) is computationally intractable. Another variant of stochastic gradient descent is obtained for sequential (time-series) data. In particular, consider data points that are gathered sequentially, one new datapoint $(\mathbf{x}^{(t)}, y^{(t)})$ at each time instant $t = 1, 2, \ldots$. For such sequential data, we can use a slight modification of the stochastic gradient descent update (5.22) to obtain an online learning method (see Sect. 4.7). This online stochastic gradient descent algorithm computes, at each time instant t,

$$\mathbf{w}^{(t+1)} := \mathbf{w}^{(t)} - \alpha_t \nabla f_{t+1}(\mathbf{w}^{(t)}). \tag{5.26}$$

5.8 Exercises

Exercise 5.1 (Use Knowledge About Problem Class) Consider the space $\mathcal{P}$ of sequences $f = (f[0], f[1], \ldots)$ that have the following properties

- they are monotone increasing, $f[r'] \geq f[r]$ for any $r' \geq r$ and $f \in \mathcal{P}$
- a change point r, where $f[r] \neq f[r+1]$ can only be at integer multiples of 100, e.g., $r = 100$ or $r = 300$.

Given some unknown function $f \in \mathcal{P}$ and starting point r_0 the problem is to find the minimum value of f as quickly as possible. We consider iterative algorithms that can query the function at some point r to obtain the values $f[r]$, $f[r-1]$ and $f[r+1]$.

Exercise 5.2 (Learning rate Schedule for stochastic gradient descent) Let us learn a linear hypothesis $h(\mathbf{x}) = \mathbf{w}^T \mathbf{x}$ using data points that arrive sequentially at discrete time instants $t = 0, 1, \ldots$. At time t, we gather a new data point $(\mathbf{x}^{(r)}, y^{(r)})$. The data points can be modelled as realizations of i.i.d. copies of a random data point $(\mathbf{x}, y)$. The probability distribution of the features $\mathbf{x}$ is a standard multivariate normal distribution $\mathcal{N}(\mathbf{0}, \mathbf{I})$. The label of a random datapoint is related to its features via $y = \overline{\mathbf{w}}^T \mathbf{x} + \varepsilon$ with some fixed but unknown "true" weight vector $\overline{\mathbf{w}}$. The additive noise $\varepsilon \sim \mathcal{N}(0, 1)$ follows a standard normal distribution. We use stochastic gradient descent to learn the weight vector $\mathbf{w}$ of a linear hypothesis,

$$\mathbf{w}^{(t+1)} = \mathbf{w}^{(t)} - \alpha_t \left((\mathbf{w}^{(t)})^T \mathbf{x}^{(t)} - y^{(t)} \right) \mathbf{x}^{(t)}. \tag{5.27}$$

with learning rate schedule $\alpha_t = \beta / t^\gamma$. Note that we compute one stochastic gradient descent iteration (5.27) for each new time step t. What conditions on the hyper-parameters β, γ ensure that $\lim_{t \to \infty} \mathbf{w}^{(t)} = \overline{\mathbf{w}}$ in distribution?

Exercise 5.3 (ImageNet.) The "ImageNet" database contains more than 10^6 images [8]. These images are labeled according to their content (e.g., does the image show a dog?). Let us assume that each image is stored as a file of at least 4 kilobytes. We want to learn a classifier that allows to predict if an image shows a dog or not. To learn this classifier we run gradient descent for logistic regression on a small computer that has 32 kilobytes memory and is connected to the internet with bandwidth of 1 Mbit/s. Therefore, for each single gradient descent update (5.4) it must essentially download all images in ImageNet. How long would such a single gradient descent update take?

Exercise 5.4 (Apple or No Apple?) Consider datapoints representing images. Each image is characterized by the RGB values (value range $0, \ldots, 255$) of 1024×1024 pixels, which we stack into a feature vector $\mathbf{x} \in \mathbb{R}^n$. We assign each image the label $y = 1$ if it shows an apple and $y = -1$ if it does not show an apple.

   We use logistic regression to learn a linear hypothesis $h(\mathbf{x}) = \mathbf{w}^T \mathbf{x}$ for classifying an image according to $\hat{y} = 1$ if $h(\mathbf{x}) \geq 0$. We use a training set of $m = 10^{10}$ labeled images which are stored in the cloud. We implement the ML method on our own laptop which is connected to the internet with a rate of at most 100 Mbps. Unfortunately we only store at most five images on our computer. How long does one single gradient descent step take at least?

Exercise 5.5 (Feature Normalization To Speed Up gradient descent) Consider the dataset with feature vectors $\mathbf{x}^{(1)} = (100, 0)^T \in \mathbb{R}^2$ and $\mathbf{x}^{(2)} = (0, 1/10)^T$ which we stack into the matrix $\mathbf{X} = (\mathbf{x}^{(1)}, \mathbf{x}^{(2)})^T$. What is the condition number of $\mathbf{X}^T\mathbf{X}$? What is the condition number of $\left(\widehat{\mathbf{X}}\right)^T\widehat{\mathbf{X}}$ with the matrix $\widehat{\mathbf{X}} = (\widehat{\mathbf{x}}^{(1)}, \widehat{\mathbf{x}}^{(2)})^T$ constructed from the normalized feature vectors $\widehat{\mathbf{x}}^{(i)}$ delivered by Algorithm 3.

References

1. I. Goodfellow, Y. Bengio, A. Courville, *Deep Learning* (MIT Press, Cambridge, 2016)
2. W. Rudin, *Principles of Mathematical Analysis*, 3rd edn. (McGraw-Hill, New York, 1976)
3. Y. Nesterov, *Introductory lectures on convex optimization*, Applied Optimization, vol. 87. (Kluwer Academic Publishers, Boston, MA, 2004)
4. S. Oymak, B. Recht, M. Soltanolkotabi, Sharp time-data tradeoffs for linear inverse problems. IEEE Trans. Inf. Theory **64**(6), 4129–4158 (2018). (June)
5. T. Hastie, R. Tibshirani, J. Friedman, *The Elements of Statistical Learning* Springer Series in Statistics. (Springer, New York, 2001)
6. A. Jung, A fixed-point of view on gradient methods for big data. Frontiers in Applied Mathematics and Statistics **3**, 1–11 (2017)
7. N. Murata, A statistical study on on-line learning, in *On-line Learning in Neural Networks*. ed. by D. Saad (Cambridge University Press, New York, 1998), pp. 63–92
8. A. Krizhevsky, I. Sutskever, G. Hinton, Imagenet classification with deep convolutional neural network, in *Neural Information Processing Systems* (NIPS, 2012)

Chapter 6
Model Validation and Selection

Chapter 4 discussed ERM as a principled approach to learning a good hypothesis out of a hypothesis space or model. ERM based methods learn a hypothesis $\hat{h} \in \mathcal{H}$ that incurs minimum average loss on some labeled data points that serve as the **training set**.[1] We refer to the average loss incurred by a hypothesis on the training set as the training error. The minimum average loss achieved by a hypothesis that solves the ERM might be referred to as the training error of the overall ML method.

ERM makes sense only if the training error of a hypothesis is a good indicator for its loss incurred on data points outside the training set. Whether the training error of a hypothesis is a reliable indicator for its performance outside the training set depends on the statistical properties of the data points and on the hypothesis space used by the ML method.

ML methods often use hypothesis spaces with a large effective dimension (see Sect. 2.2). As an example consider linear regression (see Sect. 3.1) with data points having a vast number n of features. The effective dimension of the linear hypothesis space (3.1), which is used by linear regression, is equal to the number n of features. Modern technology allows to collect a huge number of features about individual data points which implies, in turn, that the effective dimension of (3.1) is huge. Another example of high-dimensional hypothesis spaces are deep learning methods whose hypothesis spaces are constituted by all maps represented by some artificial neural network with billions of tunable weights.

A high-dimensional hypothesis space is typically very likely to contain a hypothesis that fits perfectly any given training set. Such a hypothesis achieves a very small training error but might incur a large loss when predicting the labels of data points outside the training data. The (minimum) training error achieved by a hypothesis learnt by ERM can be highly misleading. We say that a ML method, such as linear

---

[1] In statistics, the training set is also referred to as a sample.

© The Author(s), under exclusive license to Springer Nature Singapore Pte Ltd. 2022
A. Jung, *Machine Learning*, Machine Learning: Foundations, Methodologies,
and Applications, https://doi.org/10.1007/978-981-16-8193-6_6

regression using too many features, overfits the training data when it learns a hypoth-esis (e.g., via ERM) that has small training error but incurs much larger loss outside the training set.

Section 6.1 shows why linear regression will most likely overfit to the training set as soon as the number of features of a data point exceeds the size of the training set. Section 6.2 demonstrates how **to validate** a learnt hypothesis by computing its average loss on data points which are different from the training set. The data points used to validate the hypothesis are referred to as the **validation set**. When a ML method is overfitting the training set, it will learn a hypothesis whose training error is much smaller than the validation error. Thus, we can detect if a ML method overfits by comparing its training and validation errors (see Fig. 6.1).

We can use the validation error not only to detect if a ML method overfits. The validation error can also be used as a quality measure for an entire hypothesis space or model. This is analogous to the concept of a loss function that allows us to evaluate the quality of a hypothesis $h \in \mathcal{H}$. Section 6.3 shows how to do model selection based on comparing the validation errors obtained for different candidate models (hypothesis spaces).

Section 6.4 uses a simple probabilistic model for the data to study the relation between training error and the expected loss or risk of a hypothesis. The analysis of the probabilistic model reveals the interplay between the data, the hypothesis space and the resulting training and validation error of a ML method.

Section 6.5 presents the **bootstrap** method as a simulation-based alternative to the analysis of Sect. 6.4. While Sect. 6.4 assumes a particular probability distribution of data points, the bootstrap method does not require the specification of a probabil-ity distribution that underlies the data. The bootstrap method allows us to analyze statistical fluctuations in the learning process that arise from using different training sets.

As indicated in Fig. 6.1, for some ML applications, we might have a baseline level (or benchmark) for the achievable performance of ML methods. Such a base-line might be obtained from existing ML methods, human performance levels or

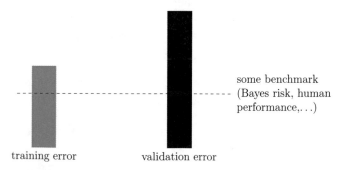

Fig. 6.1   To diagnose ML methods we compare the training with validation error. Ideally both errors are on the same level as the baseline level (if known)

from a probabilistic model (see Sect. 6.4). Section 6.6 details how the comparison between training and validation error with some benchmark error level informs possible improvements of the ML method. These improvements might be obtained by collecting more data points, using more features of data points or by changing the model (hypothesis space).

Having a baseline level also allows us to tell if a ML method already provides satisfactory results. If the training and validation error of a ML method are on the same level as the error of the theoretically optimal Bayes estimator, there is little point in modifying the ML method as it already performs (nearly) optimal.

6.1 Overfitting

We now have a closer look at the occurrence of overfitting in linear regression which is one of the ML method discussed in Sect. 3.1. Linear regression methods learn a linear hypothesis $h(\mathbf{x}) = \mathbf{w}^T \mathbf{x}$ which is parametrized by the weight vector $\mathbf{w} \in \mathbb{R}^n$. The learnt hypothesis is then used to predict the numeric label $y \in \mathbb{R}$ of a data point based on its feature vector $\mathbf{x} \in \mathbb{R}^n$.

Linear regression aims at finding a weight vector $\widehat{\mathbf{w}}$ with minimum average squared error loss incurred on a training set

$$\mathcal{D} = \left\{ \left(\mathbf{x}^{(1)}, y^{(1)} \right), \ldots, \left(\mathbf{x}^{(m)}, y^{(m)} \right) \right\}.$$

The training set consists of m data points $\left(\mathbf{x}^{(i)}, y^{(i)} \right)$, for $i = 1, \ldots, m$, with known label values $y^{(i)}$. We stack the feature vectors $\mathbf{x}^{(i)}$ and labels $y^{(i)}$ of the training data into the feature matrix $\mathbf{X} = (\mathbf{x}^{(1)}, \ldots, \mathbf{x}^{(m)})^T$ and label vector $\mathbf{y} = (y^{(1)}, \ldots, y^{(m)})^T$.

The ERM (4.13) of linear regression is solved by any weight vector $\widehat{\mathbf{w}}$ that solves (4.11). The (minimum) training error of the hypothesis $h^{(\widehat{\mathbf{w}})}$ is obtained as

$$\widehat{L}(h^{(\widehat{\mathbf{w}})} \mid \mathcal{D}) \overset{(4.4)}{=} \min_{\mathbf{w} \in \mathbb{R}^n} \widehat{L}(h^{(\mathbf{w})} | \mathcal{D})$$

$$\overset{(4.13)}{=} \|(\mathbf{I} - \mathbf{P})\mathbf{y}\|^2. \tag{6.1}$$

Here, we used the orthogonal projection matrix $\mathbf{P}$ on the linear span

$$\text{span}\{\mathbf{X}\} = \left\{ \mathbf{X}\mathbf{a} : \mathbf{a} \in \mathbb{R}^n \right\} \subseteq \mathbb{R}^m,$$

of the feature matrix $\mathbf{X} = (\mathbf{x}^{(1)}, \ldots, \mathbf{x}^{(m)})^T \in \mathbb{R}^{m \times n}$.

In many ML applications we have access to a huge number of individual features to characterize a data point. As a point in case, consider a data point which is a snapshot obtained from a modern smartphone camera. These cameras have a resolution of several megapixels. Here, we can use millions of pixel colour intensities as its features. For such applications, it is common to have more features for data points than the size of the training set,

$$n \geq m. \tag{6.2}$$

Whenever (6.2) holds, the feature vectors $\mathbf{x}^{(1)}, \ldots, \mathbf{x}^{(m)} \in \mathbb{R}^n$ of the data points in $\mathcal{D}$ are typically linearly independent. As a case in point, if the feature vectors $\mathbf{x}^{(1)}, \ldots, \mathbf{x}^{(m)} \in \mathbb{R}^n$ are realizations of i.i.d. random variables with a continuous probability distribution, these vectors are linearly independent with probability one [1].

If the feature vectors $\mathbf{x}^{(1)}, \ldots, \mathbf{x}^{(m)} \in \mathbb{R}^n$ are linearly independent, the span of the feature matrix $\mathbf{X} = (\mathbf{x}^{(1)}, \ldots, \mathbf{x}^{(m)})^T$ coincides with $\mathbb{R}^m$ which implies, in turn, $\mathbf{P} = \mathbf{I}$. Inserting $\mathbf{P} = \mathbf{I}$ into (4.13) yields

$$\widehat{L}(h^{(\widehat{\mathbf{w}})} \mid \mathcal{D}) = 0. \tag{6.3}$$

As soon as the number $m = |\mathcal{D}|$ of training data points does not exceed the number n of features that characterize data points, there is (with probability one) a linear predictor $h^{(\widehat{\mathbf{w}})}$ achieving zero training error(!).

While the hypothesis $h^{(\widehat{\mathbf{w}})}$ achieves zero training error, it will typically incur a non-zero average prediction error $y - h^{(\widehat{\mathbf{w}})}(\mathbf{x})$ on data points $(\mathbf{x}, y)$ outside the training set (see Fig. 6.2). Section 6.4 will make this statement more precise by using a probabilistic model for the data points within and outside the training set.

Note that (6.3) also applies if the features $\mathbf{x}$ and labels y of data points are completely unrelated. Consider a ML problem with data points whose labels y and features are realizations of a random variable that are statistically independent. Thus, in a very strong sense, the features $\mathbf{x}$ contain no information about the label of a data point. Nevertheless, as soon as the number of features exceeds the size of the training set, such that (6.2) holds, linear regression will learn a hypothesis with zero training error.

We can easily extend the above discussion about the occurrence of overfitting in linear regression to other methods that combine linear regression with a feature map. Polynomial regression, using data points with a single feature z, combines linear regression with the feature map $z \mapsto \mathbf{\Phi}(z) := (z^0, \ldots, z^{n-1})^T$ as discussed in Sect. 3.2.

It can be shown that whenever (6.2) holds and the features $z^{(1)}, \ldots, z^{(m)}$ of the training data are all different, the feature vectors $\mathbf{x}^{(1)} := \mathbf{\Phi}(z^{(1)}), \ldots, \mathbf{x}^{(m)} := \mathbf{\Phi}(z^{(m)})$ are linearly independent. This implies, in turn, that polynomial regression is guaranteed to find a hypothesis with zero training error whenever $m \leq n$ and the data points in the training set have different feature values.

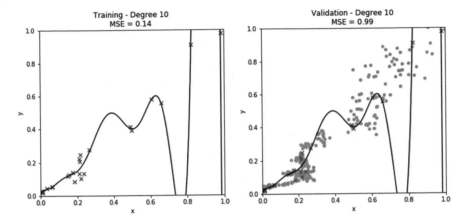

Fig. 6.2 Polynomial regression learns a polynomial map with degree $n - 1$ by minimizing its average loss on a training set (blue crosses). Using high-degree polynomials (large n) results in a small training error. However, the learnt high-degree polynomial performs poorly on data points outside the training set (orange dots)

6.2 Validation

Consider an ML method that uses ERM (4.3) to learn a hypothesis $\hat{h} \in \mathcal{H}$ out of the hypothesis space $\mathcal{H}$. The discussion in Sect. 6.1 revealed that the training error of a learnt hypothesis $\hat{h}$ can be a poor indicator for the performance of $\hat{h}$ for data points outside the training set. The hypothesis $\hat{h}$ tends to "look better" on the training set over which it has been tuned within ERM. The basic idea of validating the predictor $\hat{h}$ is simple: after learning $\hat{h}$ using ERM on a training set, compute its average loss on data points which have not been used in ERM. By **validation** we refer to the computation of the average loss on data points that have not been used in ERM.

Assume we have access to a dataset of m data points,

$$\mathcal{D} = \left\{ \left(\mathbf{x}^{(1)}, y^{(1)} \right), \ldots, \left(\mathbf{x}^{(m)}, y^{(m)} \right) \right\}.$$

Each data point is characterized by a feature vector $\mathbf{x}^{(i)}$ and a label $y^{(i)}$. Algorithm 5 outlines how to learn and validate a hypothesis $h \in \mathcal{H}$ by splitting the dataset $\mathcal{D}$ into a **training set** and a **validation set** (see Fig. 6.3).

The random shuffling in step 1 of Algorithm 5 ensures that the order of the data points has no meaning. This is important in applications where the data points are collected sequentially over time and consecutive data points might be correlated. We could avoid the shuffling step, if we construct the training set by randomly choosing a subset of size m_t instead of using the first m_t data points.

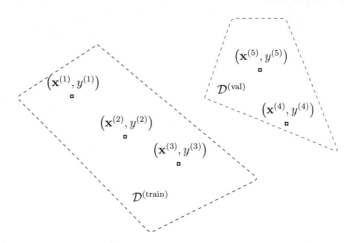

Fig. 6.3 We split the dataset $\mathcal{D}$ into two subsets, a **training set** $\mathcal{D}^{(\text{train})}$ and a **validation set** $\mathcal{D}^{(\text{val})}$. We use the training set to learn (find) the hypothesis $\hat{h}$ with minimum empirical risk $\widehat{L}(\hat{h}|\mathcal{D}^{(\text{train})})$ on the training set (4.3). We then validate $\hat{h}$ by computing its average loss $\widehat{L}(\hat{h}|\mathcal{D}^{(\text{val})})$ on the validation set $\mathcal{D}^{(\text{val})}$. The average loss $\widehat{L}(\hat{h}|\mathcal{D}^{(\text{val})})$ obtained on the validation set is the **validation error**. Note that $\hat{h}$ depends on the training set $\mathcal{D}^{(\text{train})}$ but is completely independent of the validation set $\mathcal{D}^{(\text{val})}$

Algorithm 5 Validated ERM

Input: model $\mathcal{H}$, loss function L, dataset $\mathcal{D} = \{(\mathbf{x}^{(1)}, y^{(1)}), \ldots, (\mathbf{x}^{(m)}, y^{(m)})\}$; split ratio ρ
1: randomly shuffle the data points in $\mathcal{D}$
2: create the **training set** $\mathcal{D}^{(\text{train})}$ using the first $m_t = \lceil \rho m \rceil$ data points,

$$\mathcal{D}^{(\text{train})} = \{(\mathbf{x}^{(1)}, y^{(1)}), \ldots, (\mathbf{x}^{(m_t)}, y^{(m_t)})\}.$$

3: create the **validation set** $\mathcal{D}^{(\text{val})}$ by the $m_v = m - m_t$ remaining data points,

$$\mathcal{D}^{(\text{val})} = \{(\mathbf{x}^{(m_t+1)}, y^{(m_t+1)}), \ldots, (\mathbf{x}^{(m)}, y^{(m)})\}.$$

4: learn hypothesis $\hat{h}$ via ERM on the training set,

$$\hat{h} := \underset{h \in \mathcal{H}}{\operatorname{argmin}} \widehat{L}(h|\mathcal{D}^{(\text{train})}) \tag{6.4}$$

5: compute the training error

$$E_t := \widehat{L}(\hat{h}|\mathcal{D}^{(\text{train})}) = (1/m_t) \sum_{i=1}^{m_t} L((\mathbf{x}^{(i)}, y^{(i)}), \hat{h}). \tag{6.5}$$

6: compute the **validation error**

$$E_v := \widehat{L}(\hat{h}|\mathcal{D}^{(\text{val})}) = (1/m_v) \sum_{i=m_t+1}^{m} L((\mathbf{x}^{(i)}, y^{(i)}), \hat{h}). \tag{6.6}$$

Output: learnt hypothesis $\hat{h}$, training error E_t, validation error E_v

---

6.2.1 The Size of the Validation Set

The choice of the split ratio $\rho \approx m_t/m$ in Algorithm 5 is often based on trial and error. We try out different choices for the split ratio and pick the one resulting in the smallest validation error. It is difficult to make a precise statement on how to choose the split ratio which applies broadly [2]. This difficulty stems from the fact that the optimal choice for ρ depends on the precise statistical properties of the data points.

To obtain a lower bound on the required size of the validation set, we need a probabilistic model for the data points. Let us assume that data points are realizations of i.i.d. random variables with the same probability distribution $p(\mathbf{x}, y)$. Then the validation error E_v (6.6) becomes a realization of a random variable. The expectation (or mean) $\mathbb{E}\{E_v\}$ of this RV is precisely the risk $\mathbb{E}\{L((\mathbf{x}, y), \hat{h})\}$ of $\hat{h}$ (see (4.1)).

The random validation error E_v fluctuates around its mean. We can quantify this fluctuations using the variance

$$\sigma_{E_v}^2 := \mathbb{E}\{(E_v - \mathbb{E}\{E_v\})^2\}.$$

Note that the validation error is the average of the realizations $L((\mathbf{x}^{(i)}, y^{(i)}), \hat{h})$ of i.i.d. random variables. The probability distribution of the random variable $L((\mathbf{x}, y), \hat{h})$ is determined by the probability distribution $p(\mathbf{x}, y)$, the choice of loss function and the hypothesis $\hat{h}$. In general, we do not know $p(\mathbf{x}, y)$ and, in turn, also do not know the probability distribution of $L((\mathbf{x}, y), \hat{h})$.

If we know an upper bound U on the variance of the (random) loss $L((\mathbf{x}^{(i)}, y^{(i)}), \hat{h})$, we can bound the variance of E_v as

$$\sigma_{E_v}^2 \leq U/m_v.$$

We can then, in turn, ensure that the variance $\sigma_{E_v}^2$ of the validation error E_v does not exceed a given threshold η, say $\eta = (1/100)E_t^2$, by using a validation set of size

$$m_v \geq U/\eta. \tag{6.7}$$

The lower bound (6.7) is only useful if we can determine an upper bound U on the variance of the random variable $L((\mathbf{x}, y), \hat{h})$ where $(\mathbf{x}, y)$ is a random variable with probability distribution $p(\mathbf{x}, y)$. An upper bound on the variance of $L((\mathbf{x}, y), \hat{h})$ can be derived using probability theory if we know an accurate probabilistic model $p(\mathbf{x}, y)$ for the data points. Such a probabilistic model might be provided by application-specific scientific fields such as biology or psychology. Another option is to estimate the variance of $L((\mathbf{x}, y), \hat{h})$ using the sample variance of the actual loss values $L((\mathbf{x}^{(1)}, y^{(1)}), \hat{h}), \ldots, L((\mathbf{x}^{(m)}, y^{(m)}), \hat{h})$ obtained for the dataset $\mathcal{D}$.

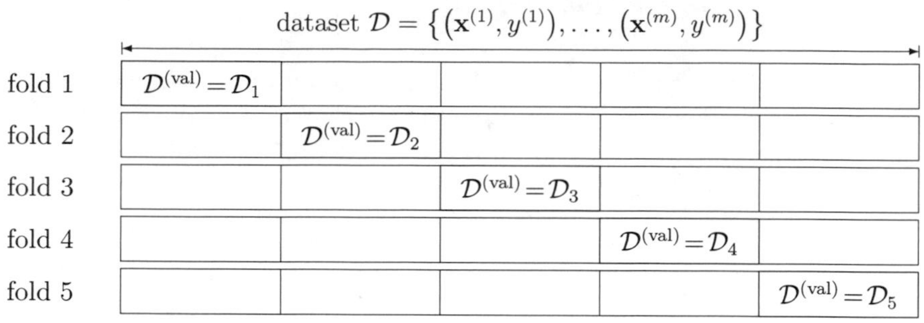

$$\text{dataset } \mathcal{D} = \left\{ \left(\mathbf{x}^{(1)}, y^{(1)} \right), \ldots, \left(\mathbf{x}^{(m)}, y^{(m)} \right) \right\}$$

fold 1	$\mathcal{D}^{(\text{val})} = \mathcal{D}_1$				
fold 2		$\mathcal{D}^{(\text{val})} = \mathcal{D}_2$			
fold 3			$\mathcal{D}^{(\text{val})} = \mathcal{D}_3$		
fold 4				$\mathcal{D}^{(\text{val})} = \mathcal{D}_4$	
fold 5					$\mathcal{D}^{(\text{val})} = \mathcal{D}_5$

Fig. 6.4 Illustration of k-fold CV for $k = 5$. We evenly partition the entire dataset $\mathcal{D}$ into $k = 5$ subsets (or folds) $\mathcal{D}_1, \ldots, \mathcal{D}_5$. We then repeat the validated ERM Algorithm 5 for $k = 5$ times. The bth repetition uses the bth fold $\mathcal{D}_b$ as the validation set and the remaining $k-1 (= 4)$ folds as the training set for ERM (4.3)

6.2.2 k-Fold Cross Validation

Algorithm 5 uses the most basic form of splitting a given dataset $\mathcal{D}$ into a training and a validation set. Many variations and extensions of this basic splitting approach have been proposed and studied (see [3] and Sect. 6.5). One very popular extension of the single split into training and validation set is known as k-**fold cross-validation** (CV) [4, Sec. 7.10]. We summarize k-fold CV in Algorithm 6 below.

Figure 6.4 illustrates the key principle behind k-fold CV which is to divide the entire dataset evenly into k subsets which are referred to as **folds**. The learning (via ERM) and validation of a hypothesis out of a given hypothesis space $\mathcal{H}$ is then repeated k times. During each repetition, we use one fold as the validation set and the remaining $k - 1$ folds as a training set. We then average the training and validation error over all repetitions.

The average (over all k folds) validation error delivered by k-fold CV is a more robust estimator for the expected loss or risk (4.1) compared to the validation error obtained from a single split. Consider a small dataset and using a single split into training and validation set. We might then be very unlucky and choose data points for the validation set which are outliers and not representative of the overall distribution of the data.

6.2.3 Imbalanced Data

The simple validation approach discussed above requires the validation set to be a good representative for the overall statistical properties of the data. This might not be the case in applications with discrete valued labels and some of the label values being very rare. We might then be interested in having a good estimate of the conditional

---

Algorithm 6 k-fold CV ERM

Input: model $\mathcal{H}$, loss function L, dataset $\mathcal{D} = \{(\mathbf{x}^{(1)}, y^{(1)}), \ldots, (\mathbf{x}^{(m)}, y^{(m)})\}$; number k of folds
1: randomly shuffle the data points in $\mathcal{D}$
2: divide the shuffled dataset $\mathcal{D}$ into k folds $\mathcal{D}_1, \ldots, \mathcal{D}_k$ of size $B = \lceil m/k \rceil$,

$$\mathcal{D}_1 = \{(\mathbf{x}^{(1)}, y^{(1)}), \ldots, (\mathbf{x}^{(B)}, y^{(B)})\}, \ldots, \mathcal{D}_k = \{(\mathbf{x}^{((k-1)B+1)}, y^{((k-1)B+1)}), \ldots, (\mathbf{x}^{(m)}, y^{(m)})\} \tag{6.8}$$

3: **for** fold index $b = 1, \ldots, k$ **do**
4:    use bth fold as the validation set $\mathcal{D}^{(\text{val})} = \mathcal{D}_b$
5:    use rest as the **training set** $\mathcal{D}^{(\text{train})} = \mathcal{D} \setminus \mathcal{D}_b$
6:    learn hypothesis $\hat{h}$ via ERM on the training set,

$$\hat{h}^{(b)} := \underset{h \in \mathcal{H}}{\text{argmin}} \, \widehat{L}(h | \mathcal{D}^{(\text{train})}) \tag{6.9}$$

7:    compute the training error

$$E_t^{(b)} := \widehat{L}(\hat{h} | \mathcal{D}^{(\text{train})}) = (1/|\mathcal{D}^{(\text{train})}|) \sum_{i \in \mathcal{D}^{(\text{train})}} L((\mathbf{x}^{(i)}, y^{(i)}), h). \tag{6.10}$$

8:    compute **validation error**

$$E_v^{(b)} := \widehat{L}(\hat{h} | \mathcal{D}^{(\text{val})}) = (1/|\mathcal{D}^{(\text{val})}|) \sum_{i \in \mathcal{D}^{(\text{val})}} L((\mathbf{x}^{(i)}, y^{(i)}), \hat{h}). \tag{6.11}$$

9: **end for**
10: compute average training and validation errors

$$E_t := (1/k) \sum_{b=1}^{k} E_t^{(b)}, \text{ and } E_v := (1/k) \sum_{b=1}^{k} E_v^{(b)}$$

11: pick a learnt hypothesis $\hat{h} := \hat{h}^{(b)}$ for some $b \in \{1, \ldots, k\}$
Output: learnt hypothesis $\hat{h}$; average training error E_t; average validation error E_v

---

risks $\mathbb{E}\{L((\mathbf{x}, y), h) | y = y'\}$ where y' is one of the rare label values. This is more than requiring a good estimate for the risk $\mathbb{E}\{L((\mathbf{x}, y), h)\}$.

Consider data points characterized by a feature vector $\mathbf{x}$ and binary label $y \in \{-1, 1\}$. Assume we aim at learning a hypothesis $h(\mathbf{x}) = \mathbf{w}^T \mathbf{x}$ to classify data points via $\hat{y} = 1$ if $h(\mathbf{x}) \geq 0$ while $\hat{y} = -1$ otherwise. The learning is based on a dataset $\mathcal{D}$ which contains only one single (!) data point with $y = -1$. If we then split the dataset into training and validation set, it is with high probability that the validation set does not include any data point with $y = -1$. This cannot happen when using k-fold CV since the single data point must be in one of the validation folds. However, even using k-fold CV for such an imbalanced dataset is problematic since we evaluate the performance of a hypothesis $h(\mathbf{x})$ using only one single data point with $y = -1$. The validation error will then be dominated by the loss of $h(\mathbf{x})$ incurred on data points with the (majority) label $y = 1$.

When learning and validating a hypothesis using imbalanced data, it might be useful to generate synthetic data points to enlarge the minority class. This can be done using data augmentation techniques which we discuss in Sect. 7.3. Another option is to use a loss function that takes the different frequency of label values into account.

Consider an imbalanced dataset of size $m = 100$, which contains 90 data points with label $y = 1$ but only 10 data points with label $y = -1$. We might then put more weight on wrong predictions obtained for the minority class (of data points with $y = -1$). This can be done by using a much larger value for the loss $L((\mathbf{x}, y = -1), h(\mathbf{x}) = 1)$ than for the loss $L((\mathbf{x}, y = 1), h(\mathbf{x}) = -1)$. Remember, the loss function is a design choice and can be freely set by the ML engineer.

6.3 Model Selection

Chapter 3 illustrated how many well-known ML methods are obtained by different combinations of a hypothesis space or model, loss function and data representation. While for many ML applications there is often a natural choice for the loss function and data representation, the right choice for the model is typically less obvious. This chapter shows how to use the validation methods of Sect. 6.2 to choose between different candidate models.

Consider data points characterized by a single numeric feature $x \in \mathbb{R}$ and numeric label $y \in \mathbb{R}$. If we suspect that the relation between feature x and label y is non-linear, we might use polynomial regression which is discussed in Sect. 3.2. Polynomial regression uses the hypothesis space $\mathcal{H}_{\text{poly}}^{(n)}$ with some maximum degree n. Different choices for the maximum degree n yield a different hypothesis space: $\mathcal{H}^{(1)} = \mathcal{H}_{\text{poly}}^{(0)}, \mathcal{H}^{(2)} = \mathcal{H}_{\text{poly}}^{(1)}, \ldots, \mathcal{H}^{(M)} = \mathcal{H}_{\text{poly}}^{(M-1)}$.

Another ML method that learns non-linear hypothesis map is Gaussian basis regression (see Sect. 3.5). Here, different choices for the variance σ and shifts μ of the Gaussian basis function (3.12) result in different hypothesis spaces. For example, $\mathcal{H}^{(1)} = \mathcal{H}_{\text{Gauss}}^{(2)}$ with $\sigma = 1$ and $\mu_1 = 1$ and $\mu_2 = 2$, $\mathcal{H}^{(2)} = \mathcal{H}_{\text{Gauss}}^{(2)}$ with $\sigma = 1/10$, $\mu_1 = 10$, $\mu_2 = 20$.

Algorithm 7 summarizes a simple method to choose between different candidate models $\mathcal{H}^{(1)}, \mathcal{H}^{(2)}, \ldots, \mathcal{H}^{(M)}$. The idea is to first learn and validate a hypothesis $\hat{h}^{(l)}$ separately for each model $\mathcal{H}^{(l)}$ using Algorithm 6. For each model $\mathcal{H}^{(l)}$, we learn the hypothesis $\hat{h}^{(l)}$ via ERM (6.4) and then compute its validation error $E_v^{(l)}$ (6.6). We then choose the hypothesis $\hat{h}^{(\hat{l})}$ from those model $\mathcal{H}^{(\hat{l})}$ which resulted in the smallest validation error $E_v^{(\hat{l})} = \min_{l=1,\ldots,M} E_v^{(l)}$.

The "work-flow" of Algorithm 7 is quite similar to the work-flow of ERM. The idea of ERM is to learn a hypothesis out of a set of different candidates (the hypothesis space). The quality of a particular hypothesis h is measured using the (average) loss incurred on some training set. We use a similar principle for model selection but on a higher level. Instead of learning a hypothesis within a hypothesis space, we choose (or learn) a hypothesis space within a set of candidate hypothesis spaces. The quality of a given hypothesis space is measured by the validation error (6.6). To determine the validation error of a hypothesis space, we first learn the hypothesis $\hat{h} \in \mathcal{H}$ via ERM (6.4) on the training set. Then, we obtain the validation error as the average loss of $\hat{h}$ on the validation set.

The final hypothesis $\hat{h}$ delivered by the model selection Algorithm 7 not only depends on the training set used in ERM (see (6.9)). This hypothesis $\hat{h}$ has also been chosen based on its validation error which is the average loss on the validation set in (6.11). Indeed, we compared this validation error with the validation errors of other models to pick the model $\mathcal{H}^{(\hat{l})}$ (see step 10) which contains $\hat{h}$. Since we used the validation error (6.11) of $\hat{h}$ to learn it, we cannot use this validation error as a good indicator for the general performance of $\hat{h}$.

To estimate the general performance of the final hypothesis $\hat{h}$ delivered by Algorithm 7 we must try it out on a test set. The test set, which is constructed in step 3 of Algorithm 7, consists of data points that have neither been used within training (6.9) or validation (6.11) of the candidate models $\mathcal{H}^{(1)}, \ldots, \mathcal{H}^{(M)}$. The average loss of the final hypothesis on the test set is referred to as the test error. The test error is computed in the step 12 of Algorithm 7.

Algorithm 7 Model Selection

Input: list of candidate models $\mathcal{H}^{(1)}, \ldots, \mathcal{H}^{(M)}$, loss function L, dataset $\mathcal{D} = \{(\mathbf{x}^{(1)}, y^{(1)}), \ldots, (\mathbf{x}^{(m)}, y^{(m)})\}$; number k of folds, test fraction ρ

1: randomly shuffle the data points in $\mathcal{D}$

2: determine size $m' := \lceil \rho m \rceil$ of test set

3: construct **test set**

$$\mathcal{D}^{(\text{test})} = \{(\mathbf{x}^{(1)}, y^{(1)}), \ldots, (\mathbf{x}^{(m')}, y^{(m')})\}$$

4: construct the set used for training and validation,

$$\mathcal{D}^{(\text{trainval})} = \{(\mathbf{x}^{(m'+1)}, y^{(m'+1)}), \ldots, (\mathbf{x}^{(m)}, y^{(m)})\}$$

5: **for** model index $l = 1, \ldots, M$ **do**

6:     run Algorithm 6 using $\mathcal{H} = \mathcal{H}^{(l)}$, dataset $\mathcal{D} = \mathcal{D}^{(\text{trainval})}$, loss function L and k folds

7:     Algorithm 6 delivers hypothesis $\hat{h}$ and validation error E_v

8:     store learnt hypothesis $\hat{h}^{(l)} := \hat{h}$ and validation error $E_v^{(l)} := E_v$

9: **end for**

10: pick model $\mathcal{H}^{(\hat{l})}$ with minimum validation error $E_v^{(\hat{l})} = \min_{l=1,\ldots,M} E_v^{(l)}$

11: define optimal hypothesis $\hat{h} = \hat{h}^{(\hat{l})}$

12: compute **test error**

$$E^{(\text{test})} := \widehat{L}(\hat{h}|\mathcal{D}^{(\text{test})}) = (1/|\mathcal{D}^{(\text{test})}|) \sum_{i \in \mathcal{D}^{(\text{test})}} L((\mathbf{x}^{(i)}, y^{(i)}), \hat{h}). \tag{6.12}$$

Output: hypothesis $\hat{h}$; training error $E_t^{(\hat{l})}$; validation error $E_v^{(\hat{l})}$, test error $E^{(\text{test})}$.

Sometimes it is beneficial to use different loss functions for the training and the validation of a hypothesis. As an example, consider the ML methods logistic regression and the support vector machine which have been discussed in Sects. 3.6

and 3.7, respectively. Both methods use the same model which is the space of linear hypothesis maps $h(\mathbf{x}) = \mathbf{w}^T \mathbf{x}$. The main difference between these two methods is the choice for the loss function used to measure the quality of a hypothesis. Logistic regression minimizes the (average) logistic loss (2.12) on the training set to learn the hypothesis $h^{(1)}(\mathbf{x}) = \left(\mathbf{w}^{(1)}\right)^T \mathbf{x}$ with a weight vector $\mathbf{w}^{(1)}$. The support vector machine instead minimizes the (average) hinge loss (2.11) on the training set to learn the hypothesis $h^{(2)}(\mathbf{x}) = \left(\mathbf{w}^{(2)}\right)^T \mathbf{x}$ with a weight vector $\mathbf{w}^{(2)}$. It would be difficult to compare the hypotheses $h^{(1)}(\mathbf{x})$ and $h^{(2)}(\mathbf{x})$ using different loss functions to compute their validation errors. For a comparison, we could instead compute the validation errors for $h^{(1)}(\mathbf{x})$ and $h^{(2)}(\mathbf{x})$ using the average 0/1 loss (2.9) ("accuracy").

Algorithm 7 requires as one of its inputs a given list of candidate models. The longer this list, the more computation is required from Algorithm 7. Sometimes it is possible to prune the list of candidate models by removing models that are very unlikely to have minimum validation error.

Consider polynomial regression which uses as the model the space $\mathcal{H}_{poly}^{(r)}$ of polynomials with maximum degree r (see (3.4)). For $r = 1$, $\mathcal{H}_{poly}^{(r)}$ is the space of polynomials with maximum degree one (which are linear maps), $h(x) = w_2 x + w_1$. For $r = 2$, $\mathcal{H}_{poly}^{(r)}$ is the space of polynomials with maximum degree two, $h(x) = w_3 x^2 + w_2 x + w_1$.

The polynomial degree r parametrizes a nested set of models,

$$\mathcal{H}_{poly}^{(1)} \subset \mathcal{H}_{poly}^{(2)} \subset \ldots.$$

For each degree r, we learn a hypothesis $h^{(r)} \in \mathcal{H}_{poly}^{(r)}$ with minimum average loss (training error) $E_t^{(r)}$ on a training set (see (6.5)). To validate the learnt hypothesis $h^{(r)}$, we compute its average loss (validation error) $E_v^{(r)}$ on a validation set (see (6.6)).

Figure 6.5 depicts the typical dependency of the training and validation errors on the polynomial degree r. The training error $E_t^{(r)}$ decreases monotonically with increasing degree r. To understand why this is the case, consider the two specific choices $r = 3$ and $r = 5$ with corresponding models $\mathcal{H}_{poly}^{(3)}$ and $\mathcal{H}_{poly}^{(5)}$. Note that $\mathcal{H}_{poly}^{(3)} \subset \mathcal{H}_{poly}^{(5)}$ since any polynomial with degree not exceeding 3 is also a polynomial with degree not exceeding 5. Therefore, the training error (6.5) obtained when minimizing over the larger model $\mathcal{H}_{poly}^{(5)}$ can only decrease but never increase compared to (6.5) using the smaller model $\mathcal{H}_{poly}^{(3)}$

Figure 6.5 indicates that the validation error $E_v^{(r)}$ (see (6.6)) behaves very different compared to the training error $E_t^{(r)}$. Starting with degree $r = 0$, the validation error first decreases with increasing degree r. As soon as the degree r is increased beyond a critical value, the validation error starts to increase with increasing r. For very large values of r, the training error becomes almost negligible while the validation error becomes very large. In this regime, polynomial regression overfits the training set.

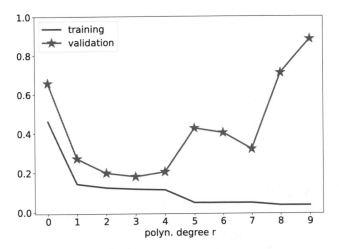

Fig. 6.5 The training error and validation error obtained from polynomial regression using different values r for the maximum polynomial degree

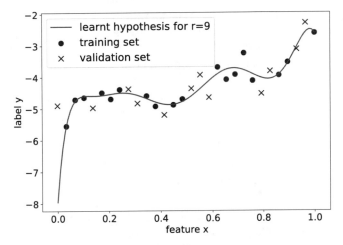

Fig. 6.6 A hypothesis $\hat{h}$ which is a polynomial with degree not larger than $r = 9$. The hypothesis has been learnt by minimizing the average loss on the training set. Note the fast rate of the change of $\hat{h}$ for feature values $x \approx 0$

Figure 6.6 illustrates the overfitting of polynomial regression when using a maximum degree that is too large. In particular, Fig. 6.6 depicts a learnt hypothesis which is a degree 9 polynomial that fits very well the training set, resulting in a very small training error. To achieve this low training error the resulting polynomial has an unreasonable high rate of change for feature values $x \approx 0$. This results in large prediction errors for validation data points with feature values $x \approx 0$.

6.4   A Probabilistic Analysis of Generalization

More Data Beats Clever Algorithms?; More Data Beats Clever Feature Selection?
A key challenge in ML is to ensure that a hypothesis that predicts well the labels on a training set (which has been used to learn that hypothesis) will also predict well the labels of data points outside the training set. We say that a ML method generalizes if a small loss on the training set implies small loss on data points outside the training set.

To study the generalization of linear regression methods (see Sect. 3.1), we will use a probabilistic model for the data. We interpret data points as i.i.d. realizations of random variables that have the same distribution as a random data point $\mathbf{z} = (\mathbf{x}, y)$. The random feature vector $\mathbf{x}$ is assumed to have zero mean and covariance being the identity matrix, i.e., $\mathbf{x} \sim \mathcal{N}(\mathbf{0}, \mathbf{I})$. The label y of a random data point is related to its features $\mathbf{x}$ via a **linear Gaussian model**

$$y = \bar{\mathbf{w}}^T \mathbf{x} + \varepsilon, \text{ with noise } \varepsilon \sim \mathcal{N}(0, \sigma^2). \tag{6.13}$$

We assume the noise variance σ^2 fixed and known. This is a simplifying assumption as in practice, we would need to estimate the noise variance from data [5]. Note that, within our probabilistic model, the error component ε in (6.13) is intrinsic to the data and cannot be overcome by any ML method. We highlight that the probabilistic model for the observed data points is just a modelling assumption. This assumption allows us to study some fundamental behaviour of ML methods. There are principled methods ("tests") that allow to determine if a given dataset can be accurately modelled using (6.13) [6].

We predict the label y from the features $\mathbf{x}$ using a linear hypothesis $h(\mathbf{x})$ that depends only on the first r features $x_1, \ldots, x_r$. Thus, we use the hypothesis space

$$\mathcal{H}^{(r)} = \{h^{(\mathbf{w})}(\mathbf{x}) = (\mathbf{w}^T, \mathbf{0})\mathbf{x} \text{ with } \mathbf{w} \in \mathbb{R}^r\}. \tag{6.14}$$

The design parameter r determines the size of the hypothesis space $\mathcal{H}^{(r)}$ and, in turn, the computational complexity of learning the optimal hypothesis in $\mathcal{H}^{(r)}$.

For $r < n$, the hypothesis space $\mathcal{H}^{(r)}$ is a proper subset of the space of linear predictors (2.4) used within linear regression (see Sect. 3.1). Note that each element $h^{(\mathbf{w})} \in \mathcal{H}^{(r)}$ corresponds to a particular choice of the weight vector $\mathbf{w} \in \mathbb{R}^r$.

The quality of a particular predictor $h^{(\mathbf{w})} \in \mathcal{H}^{(r)}$ is measured via the mean squared error $\widehat{L}(h^{(\mathbf{w})} \mid \mathcal{D}^{(\text{train})})$ incurred on the labeled training set $\mathcal{D}^{(\text{train})} = \{\mathbf{x}^{(i)}, y^{(i)}\}_{i=1}^{m_t}$. Within our toy model (see (6.13), (6.15) and (6.16)), the training data points $(\mathbf{x}^{(i)}, y^{(i)})$ are i.i.d. copies of the data point $\mathbf{z} = (\mathbf{x}, y)$.

The data points in the training dataset and any other data points outside the training set are statistically independent. However, the training data points $(\mathbf{x}^{(i)}, y^{(i)})$ and any other data point $(\mathbf{x}, y)$ are drawn from the same probability distribution, which is a multivariate normal distribution,

$$\mathbf{x}, \mathbf{x}^{(i)} \text{ i.i.d. with } \mathbf{x}, \mathbf{x}^{(i)} \sim \mathcal{N}(\mathbf{0}, \mathbf{I}) \tag{6.15}$$

and the labels $y^{(i)}$, y are obtained as

$$y^{(i)} = \bar{\mathbf{w}}^T \mathbf{x}^{(i)} + \varepsilon^{(i)}, \text{ and } y = \bar{\mathbf{w}}^T \mathbf{x} + \varepsilon \tag{6.16}$$

with i.i.d. noise $\varepsilon, \varepsilon^{(i)} \sim \mathcal{N}(0, \sigma^2)$.

As discussed in Chap. 4, the training error $\widehat{L}(h^{(\mathbf{w})} \mid \mathcal{D}^{(\text{train})})$ is minimized by the predictor $h^{(\widehat{\mathbf{w}})}(\mathbf{x}) = \widehat{\mathbf{w}}^T \mathbf{I}_{r \times n} \mathbf{x}$, with weight vector

$$\widehat{\mathbf{w}} = (\mathbf{X}_r^T \mathbf{X}_r)^{-1} \mathbf{X}_r^T \mathbf{y} \tag{6.17}$$

with feature matrix $\mathbf{X}_r$ and label vector $\mathbf{y}$ defined as

$$\mathbf{X}_r = (\mathbf{x}^{(1)}, \dots, \mathbf{x}^{(m_t)})^T \mathbf{I}_{n \times r} \in \mathbb{R}^{m_t \times r}, \text{ and}$$
$$\mathbf{y} = \left(y^{(1)}, \dots, y^{(m_t)} \right)^T \in \mathbb{R}^{m_t}. \tag{6.18}$$

It will be convenient to tolerate a slight abuse of notation and denote both, the length-r vector (6.17) as well as the zero padded length-n vector $(\widehat{\mathbf{w}}^T, \mathbf{0})^T$, by $\widehat{\mathbf{w}}$. This allows us to write

$$h^{(\widehat{\mathbf{w}})}(\mathbf{x}) = \widehat{\mathbf{w}}^T \mathbf{x}. \tag{6.19}$$

We highlight that the formula (6.17) for the optimal weight vector $\widehat{\mathbf{w}}$ is only valid if the matrix $\mathbf{X}_r^T \mathbf{X}_r$ is invertible. However, it can be shown that within our toy model (see (6.15)), this is true with probability one whenever $m_t \geq r$. In what follows, we will consider the case of having more training samples than the dimension of the hypothesis space, i.e., $m_t > r$ such that the formula (6.17) is valid (with probability one). The case $m_t \leq r$ will be studied in Chap. 7.

The optimal weight vector $\widehat{\mathbf{w}}$ (see (6.17)) depends on the training data $\mathcal{D}^{(\text{train})}$ via the feature matrix $\mathbf{X}_r$ and label vector $\mathbf{y}$ (see (6.18)). Therefore, since we model the training data as random, the weight vector $\widehat{\mathbf{w}}$ (6.17) is a random quantity. For each different realization of the training dataset, we obtain a different realization of the optimal weight $\widehat{\mathbf{w}}$.

The probabilistic model (6.13) relates the features $\mathbf{x}$ of a data point to its label y via some (unknown) true weight vector $\bar{\mathbf{w}}$. Intuitively, the best linear hypothesis would be $h(\mathbf{x}) = \widehat{\mathbf{w}}^T \mathbf{x}$ with weight vector $\widehat{\mathbf{w}} = \bar{\mathbf{w}}$. However, in general this will not be achievable since we have to compute $\widehat{\mathbf{w}}$ based on the features $\mathbf{x}^{(i)}$ and noisy labels $y^{(i)}$ of the data points in the training dataset $\mathcal{D}$.

In general, learning the weights of a linear hypothesis by ERM (4.5) results in a non-zero **estimation error**

$$\Delta \mathbf{w} := \widehat{\mathbf{w}} - \bar{\mathbf{w}}. \tag{6.20}$$

The estimation error (6.20) is the realization of a random variable since the learnt weight vector $\widehat{\mathbf{w}}$ (see (6.17)) is itself a realization of a random variable.

Bias and Variance. As we will see below, the prediction quality achieved by $h^{(\widehat{\mathbf{w}})}$ depends crucially on the mean squared estimation error

$$E_{\text{est}} := \mathbb{E}\{\|\Delta\mathbf{w}\|_2^2\} = \mathbb{E}\{\|\widehat{\mathbf{w}} - \bar{\mathbf{w}}\|_2^2\}. \tag{6.21}$$

We can decompose the MSE $\mathcal{E}_{\text{est}}$ into two components. The first component is the **bias** which characterizes the average behaviour, over all different realizations of training sets, of the learnt hypothesis. The second component is the **variance** which quantifies the amount of random fluctuations of the hypothesis obtained from ERM applied to different realizations of the training set. Both components depend on the model complexity parameter r.

It is not too difficult to show that

$$E_{\text{est}} = \underbrace{\|\bar{\mathbf{w}} - \mathbb{E}\{\widehat{\mathbf{w}}\}\|_2^2}_{\text{"bias"}\,B^2} + \underbrace{\mathbb{E}\|\widehat{\mathbf{w}} - \mathbb{E}\{\widehat{\mathbf{w}}\}\|_2^2}_{\text{"variance"}\,V} \tag{6.22}$$

The bias term in (6.22), which can be computed as

$$B^2 = \|\bar{\mathbf{w}} - \mathbb{E}\{\widehat{\mathbf{w}}\}\|_2^2 = \sum_{l=r+1}^{n} \bar{w}_l^2, \tag{6.23}$$

measures the distance between the "true hypothesis" $h^{(\bar{\mathbf{w}})}(\mathbf{x}) = \bar{\mathbf{w}}^T\mathbf{x}$ and the hypothesis space $\mathcal{H}^{(r)}$ (see (6.14)) of the linear regression problem.

The bias (6.23) is zero if $\bar{w}_l = 0$ for any index $l = r + 1, \ldots, n$, or equivalently if $h^{(\bar{\mathbf{w}})} \in \mathcal{H}^{(r)}$. We can ensure that for every possible true weight vector $\bar{\mathbf{w}}$ in (6.13) only if we use the hypothesis space $\mathcal{H}^{(r)}$ with $r = n$.

When using the model $\mathcal{H}^{(r)}$ with $r < n$, we cannot guarantee a zero bias term since we have no access to the true underlying weight vector $\bar{\mathbf{w}}$ in (6.13). In general, the bias term decreases with an increasing model size r (see Fig. 6.7). We highlight that the bias term does not depend on the variance σ^2 of the noise ε in our toy model (6.13).

Let us now consider the variance term in (6.22). Using the properties of our toy model (see (6.13), (6.15) and (6.16))

$$V = \mathbb{E}\{\|\widehat{\mathbf{w}} - \mathbb{E}\{\widehat{\mathbf{w}}\}\|_2^2\} = \sigma^2 \text{tr}\{\mathbb{E}\{(\mathbf{X}_r^T\mathbf{X}_r)^{-1}\}\}. \tag{6.24}$$

By (6.15), the matrix $(\mathbf{X}_r^T\mathbf{X}_r)^{-1}$ is random and distributed according to an **inverse Wishart distribution** [7]. For $m_t > r + 1$, its expectation is given as

$$\mathbb{E}\{(\mathbf{X}_r^T\mathbf{X}_r)^{-1}\} = 1/(m_t - r - 1)\mathbf{I}_{r\times r}. \tag{6.25}$$

By inserting (6.25) and $\text{tr}\{\mathbf{I}_{r\times r}\} = r$ into (6.24),

$$V = \mathbb{E}\{\|\widehat{\mathbf{w}} - \mathbb{E}\{\widehat{\mathbf{w}}\}\|_2^2\} = \sigma^2 r/(m_t - r - 1). \tag{6.26}$$

As indicated by (6.26), the variance term increases with increasing model complexity r (see Fig. 6.7). This behaviour is in stark contrast to the bias term which decreases with increasing r. The opposite dependency of bias and variance on the

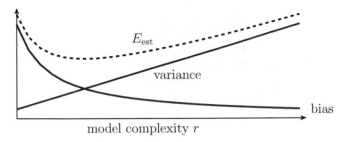

Fig. 6.7 The estimation error E_{est} incurred by linear regression can be decomposed into a bias term B^2 and a variance term V (see (6.22)). These two components depend on the model complexity r in an opposite manner resulting in a bias-variance trade-off

model complexity is known as the **bias-variance trade-off**. Thus, the choice of model complexity r (see (6.14)) has to balance between a small variance and a small bias.

Generalization. Consider the linear hypothesis $h(\mathbf{x}) = \widehat{\mathbf{w}}^T \mathbf{x}$ with the weight vector (6.17) which results in a minimum training error. We would like this predictor to generalize well to data points which are different from the training set. This generalization capability can be quantified by the expected loss or risk

$$
\begin{aligned}
E_{\text{pred}} &= \mathbb{E}\{(y - \hat{y})^2\} \\
&\overset{(6.13)}{=} \mathbb{E}\{\Delta \mathbf{w}^T \mathbf{x}\mathbf{x}^T \Delta \mathbf{w}\} + \sigma^2 \\
&\overset{(a)}{=} \mathbb{E}\{\mathbb{E}\{\Delta \mathbf{w}^T \mathbf{x}\mathbf{x}^T \Delta \mathbf{w} \mid \mathcal{D}\}\} + \sigma^2 \\
&\overset{(b)}{=} \mathbb{E}\{\Delta \mathbf{w}^T \Delta \mathbf{w}\} + \sigma^2 \\
&\overset{(6.20),(6.21)}{=} E_{\text{est}} + \sigma^2 \\
&\overset{(6.22)}{=} B^2 + V + \sigma^2.
\end{aligned} \tag{6.27}
$$

Step (a) uses the law of total expectation [8] and step (b) uses that, conditioned on the dataset $\mathcal{D}$, the feature vector $\mathbf{x}$ of a new data point is a random vector with zero mean and a covariance matrix $\mathbb{E}\{\mathbf{x}\mathbf{x}^T\} = \mathbf{I}$ (see (6.15)).

According to (6.27), the average (expected) prediction error E_{pred} is the sum of three components: (i) the bias B^2, (ii) the variance V and (iii) the noise variance σ^2. Figure 6.7 illustrates the typical dependency of the bias and variance on the model, which is parametrized by the model complexity r. which also coincides with our notion of effective model dimension (see Sect. 2.2).

The bias and variance, whose sum is the estimation error E_{est}, can be influenced by varying the model complexity r which is a design parameter. The noise variance σ^2 is the intrinsic accuracy limit of our toy model (6.13) and is not under the control of the ML engineer. It is impossible for any ML method - no matter how advanced it is - to achieve, on average, a prediction error smaller than the noise variance σ^2.

We finally highlight that our analysis of bias (6.23), variance (6.26) and the average prediction error (6.27) only applies if the observed data points are well modelled as realizations of random vectors according to (6.13), (6.15) and (6.16). The usefulness of this model for the data arising in a particular application has to be verified in practice by statistical model validation techniques [9, 10].

The qualitative behaviour of estimation error in Fig. 6.7 depends on the definition for the model complexity. Our concept of effective dimension (see Sect. 2.2) coincides with most other notions of model complexity for the linear hypothesis space (6.14). However, for more complicated models such as deep nets it is often not obvious how the model complexity is related to more tangible quantities such as total number of tunable weights or artificial neurons.

In general, the model complexity or effective model dimension is not directly proportional to number of tunable weights but also depends on the specific learning algorithm such as stochastic gradient descent. Therefore, for deep nets, if we would plot estimation error against number of tunable weights we might observe a behaviour of estimation error fundamentally different from the shape in Fig. 6.7. One example for such un-intuitive behaviour is known as "double descent phenomena" [11].

An alternative approach for analyzing bias, variance and average prediction error of linear regression is to use simulations. Here, we generate a number of i.i.d. copies of the observed data points by some random number generator [12]. Using these i.i.d. copies, we can replace exact computations (expectations) by empirical approximations (sample averages).

6.5   The Bootstrap

basic idea of bootstrap: use empirical distribution (histogram) of data points as their probability distribution; we can then sample any amount of new data points from that distribution (using sampling with replacement)

Consider learning a hypothesis $\hat{h} \in \mathcal{H}$ by minimizing the average loss incurred on a dataset $\mathcal{D} = \{(\mathbf{x}^{(1)}, y^{(1)}), \ldots, (\mathbf{x}^{(m)}, y^{(m)})\}$. The data points $(\mathbf{x}^{(i))}, y^{(i)})$ are modelled as realizations of i.i.d. random variables. Let use denote the (common) probability distribution of these random variables by $p(\mathbf{x}, y)$.

If we interpret the data points $(\mathbf{x}^{(i))}, y^{(i)})$ as realizations of random variables, also the learnt hypothesis $\hat{h}$ is a realization of a random variable. Indeed, the hypothesis $\hat{h}$ is obtained by solving an optimization problem (4.3) that involves realizations of random variables. The bootstrap is a method for estimating (parameters of) the probability distribution $p(\hat{h})$ [4].

Section 6.4 used a probabilistic model for data points to derive analytically (some parameters of) the probability distribution $p(\hat{h})$. While the analysis in Sect. 6.4 only applies to the specific probabilistic model (6.15), (6.16), the bootstrap can be used for data points drawn from an arbitrary probability distribution.

The core idea behind the bootstrap is to use the empirical distribution or histogram $\hat{p}(\mathbf{z})$ of the available data points $\mathcal{D}$ to generate B new datasets $\mathcal{D}^{(1)}, \ldots$. Each dataset is constructed such that is has the same size as the original dataset $\mathcal{D}$. For each dataset $\mathcal{D}^{(b)}$, we solve a separate ERM (4.3) to obtain the hypothesis $\hat{h}^{(b)}$. The hypothesis $\hat{h}^{(b)}$ is a realization of a random variable whose distribution is determined by the empirical distribution $\hat{p}(\mathbf{z})$ as well as the hypothesis space and the loss function used in the ERM (4.3).

6.6 Diagnosing ML

compare training, validation and benchmark error. benchmark can be a Bayes risk when using a probabilistic model (such as the i.i.d. assumption), or human performance or risk of some other ML methods ("experts" in regret framework)

In what follows, we assume that data points can (to a good approximation) be interpreted as realizations of i.i.d. random variables (see Sect. 2.1.4). This "i.i.d. assumption" underlies ERM (4.3) as the guiding principle for learning a hypothesis with small risk (4.1). This assumption also motivates to use the average loss (6.6) on a validation set as an estimate for the risk.

Consider a ML method which uses Algorithm 5 (or Algorithm 6) to learn and validate the hypothesis $\hat{h} \in \mathcal{H}$. Besides the learnt hypothesis $\hat{h}$, these algorithms also deliver the training error E_t and the validation error E_v. As we will see shortly, we can diagnose ML methods to some extend just by comparing training with validation errors. This diagnosis is further enabled if we know a benchmark (or reference) error level $E^{(\text{ref})}$.

One important source of a benchmark error level $E^{(\text{ref})}$ are probabilistic models for the data points (see Sect. 6.4). Given a probabilistic model, which specifies the probability distribution $p(\mathbf{x}, y)$ of the features and label of data points, we can compute the minimum achievable expected loss or risk (4.1). Indeed, the minimum achievable risk is precisely the expected loss of the Bayes estimator $\hat{h}(x)$ of the label y, given the features $\mathbf{x}$ of a data point. The Bayes estimator $\hat{h}(x)$ is fully determined by the probability distribution $p(\mathbf{x}, y)$ of the features and label of a (random) data point [13, Chap. 4].

Example. Let us derive the minimum achievable risk for data points with single numeric feature and label and being realizations of a Gaussian random vector $\mathbf{z} \sim \mathcal{N}(0, \mathbf{C})$. Here, the optimal estimator of the label y given the feature x is the conditional expectation of the (unobserved) label y given the (observed) feature x. The resulting MSE is equal to the posterior variance of y, given x which is given by the $K_{y,y}^{-1}$ with the entry $K_{y,y}$ of the precision matrix $\mathbf{K} = \mathbf{C}^{-1}$.

A further potential source for a benchmark error level $E^{(\text{ref})}$ is another ML method. This other ML method might be computationally too expensive to be used for a ML application. However, we could still use its error level measured in illustrative test scenarios as a benchmark.

Finally, a benchmark can be obtained from the performance of human experts. If we want to develop a ML method that detects certain type of skin cancers from images of the skin, a benchmark might be the current classification accuracy achieved by experienced dermatologists [14].

We can diagnose a ML method by comparing the training error E_t with the validation error E_v and (if available) the benchmark $E^{(\text{ref})}$.

- $E_t \approx E_v \approx E^{(\text{ref})}$: The training error is on the same level as the validation error and the benchmark error. There is not much to improve here since the validation error is already on the desired error level. Moreover, the training error is not much smaller than the validation error which indicates that there is no overfitting. It seems we have obtained a ML method that achieves the benchmark error level.

- $E_v \gg E_t$: The validation error is significantly larger than the training error. It seems that the ERM (4.3) results in a hypothesis $\hat{h}$ that overfits the training set. The loss incurred by $\hat{h}$ on data points outside the training set, such as those in the validation set, is significantly worse. This is an indicator for overfitting which can be addressed either by reducing the effective size of the hypothesis space or by increasing the effective number of training data points. To reduce the effective hypothesis space we can either use a smaller hypothesis space, e.g., using fewer features in a linear model (3.1), using smaller maximum depth of decision trees (Sect. 3.10) or by using a smaller artificial neural network (Sect. 3.11). Another way to reduce the effective size of a hypothesis space is to use regularization techniques from Chap. 7.

- $E_t \approx E_v \gg E^{(\text{ref})}$: The training error is on the same level as the validation error and both are significantly larger than the benchmark error. Since the training error is not much smaller than the validation error, the learnt hypothesis seems to not over-fit the training data. However, the training error achieved by the learnt hypothesis is significantly larger than the benchmark error level. There can be several reasons for this to happen. First, it might be that the hypothesis space used by the ML method is too small, i.e., it does not include a hypothesis that provides a good approximation for the relation between features and label of a data point. The remedy for this situation is to use a larger hypothesis space, e.g., by including more features in a linear model, using higher polynomial degrees in polynomial regression, using deeper decision trees or having larger artificial neural networks (deep nets). Another reason for the training error being too large is that the optimization algorithms used to solve ERM (4.3) is not working properly. When using gradient based optimization (see Sect. 5.4) to solve ERM, one reason for $E_t \gg E^{(\text{ref})}$ could be that the step size α in the gradient descent step (5.4) is chosen too small or too large (see Fig. 5.3-(b)). This can be solved by adjusting the step-size by trying out several different values and using the one resulting in minimum training and validation error. Another option is to use some probabilistic model for data points derive optimal values for the step-size based on such a model (see Sect. 6.4).

- $E_v \gg E_t$: The validation error is significantly larger than the training error. The idea of ERM (4.3) is to approximate the risk (4.1) of a hypothesis by its average loss on a training set $\mathcal{D} = \{(\mathbf{x}^{(i)}, y^{(i)})\}_{i=1}^{m}$. The mathematical underpinning for this approximation is the law of large numbers which characterizes the average of i.i.d. random variables. The quality of this approximation requires two conditions. First, the data points used for computing the average loss should be such that they would be typically obtained as realizations of i.i.d. random variables with a common probability distribution. Second the number of data points used for computing the average loss must be sufficiently large. Thus, if data points cannot be modelled as realizations of i.i.d. random variables or if the size of the training or validation set is too small, the interpretation (comparison) of validation and training errors is difficult. In particular, it might then be that the validation set consists of data points for which any predictor incurs small average loss. Here, we might try to increase training and validation sets by collecting more labeled data points or using data augmentation (see Sect. 7.3). If we already have quite large training and validation sets, one should verify if data points conform to the i.i.d. assumption that is required for the ERM to deliver a hypothesis with small risk. There are principled methods to test if an i.i.d. assumption is satisfied (see [15] and references therein).

6.7 Exercises

Exercise 6.1 (Validation Set Size.) Consider a linear regression problem with data points characterized by a scalar feature and a numeric label. Assume data points are realizations of i.i.d. Gaussian random variables with zero-mean and covariance matrix $\mathbf{C}$. How many data points do we need to include in the validation set such that with probability of at least 0.8 the validation error does not deviate by more than 20 percent from the expected loss or risk?

Exercise 6.2 (Validation Error Smaller Than training error?) Linear regression determines a linear hypothesis map by minimizing the average squared error on a training set. The resulting linear predictor is then validated on a validation set which is different from the training set. Can you construct a training set and validation set such that the validation error is strictly smaller than the training error?

Exercise 6.3 (When is Validation Set Too Small?) The usefulness of the validation error as an indicator for the performance of a hypothesis depends on the size of the validation set. Experiment with different ML methods and datasets to find out the minimum required size for the validation set.

References

1. R. Muirhead, *Aspects of Multivariate Statistical Theory* (Wiley, New York, 1982)
2. J. Larsen, C. Goutte, On optimal data split for generalization estimation and model selection, in *IEEE Workshop on Neural Networks for Signal Process* (IEEE, New York, 1999)
3. B. Efron, R. Tibshirani, Improvements on cross-validation: The 632+ bootstrap method. J. Am. Stat. Assoc. **92**(438), 548–560 (1997)
4. T. Hastie, R. Tibshirani, J. Friedman, *The Elements of Statistical Learning* Springer Series in Statistics. (Springer, New York, 2001)
5. I. Cohen, B. Berdugo, Noise estimation by minima controlled recursive averaging for robust speech enhancement. IEEE Sig. Proc. Lett. **9**(1), 12–15 (2002). (Jan.)
6. P. Huber, Approximate models, in *Goodness-of-Fit Tests and Model Validity*. ed. by C. Huber-Carol, N. Balakrishnan, M. Nikulin, M. Mesbah. Statistics for Industry and Technology. (Birkhäuser, Boston, MA, 2002)
7. K.V. Mardia, J.T. Kent, J.M. Bibby, *Multivariate Analysis* (Academic Press, London, 1979)
8. P. Billingsley, *Probability and Measure*, 3rd edn. (Wiley, New York, 1995)
9. K. Young, Bayesian diagnostics for checking assumptions of normality. J. Stat. Comput. Simul. **47**(3–4), 167–180 (1993)
10. O. Vasicek, A test for normality based on sample entropy. J. Roy. Stat. Soc.: Ser. B (Methodol.) **38**(1), 54–59 (1976)
11. M. Belkin, D. Hsu, S. Ma, S. Mandal, Reconciling modern machine-learning practice and the classical bias–variance trade-off. Proc. Natl. Acad. Sci. **116**(32), 15849–15854 (2019)
12. C. Andrieu, N. de Freitas, A. Doucet, M.I. Jordan, An introduction to MCMC for machine learning. Mach. Learn. **50**(1–2), 5–43 (2003)
13. E.L. Lehmann, G. Casella, *Theory of Point Estimation*, 2nd edn. (Springer, New York, 1998)
14. A. Esteva, B. Kuprel, R.A. Novoa, J. Ko, S.M. Swetter, H.M. Blau, S. Thrun, Dermatologist-level classification of skin cancer with deep neural networks. Nature **542**, 115–118 (2017)
15. H. Lütkepohl, *New Introduction to Multiple Time Series Analysis* (Springer, New York, 2005)

Chapter 7
Regularization

Keywords Data Augmentation · Robustness · Semi-Supervised Learning · Transfer Learning · Multitask Learning

Many ML methods use the principle of ERM (see Chap. 4) to learn a hypothesis out of a hypothesis space by minimizing the average loss (training error) on a set of labeled data points (training set). Using ERM as a guiding principle for ML methods makes sense only if the training error is a good indicator for its loss incurred outside the training set.

Figure 7.1 illustrates a typical scenario for a modern ML method which uses a large hypothesis space. This large hypothesis space includes highly non-linear maps which can perfectly resemble any dataset of modest size. However, there might be non-linear maps for which a small training error does not guarantee accurate predictions for the labels of data points outside the training set.

Chapter 6 discussed validation techniques to verify if a hypothesis with small training error will predict also well the labels of data points outside the training set. These validation techniques, including Algorithms 5 and 6, probe the hypothesis $\hat{h} \in \mathcal{H}$ delivered by ERM on a validation set. The validation set consists of data points which have not been used for the training set of ERM (4.3). The validation error, which is the average loss of the hypothesis on the data points in the validation set, serves as an estimate for the average error or risk (4.1) of the hypothesis $\hat{h}$.

This chapter discusses regularization as an alternative to validation techniques. In contrast to validation, regularization techniques do not require having a separate validation set which is not used for the ERM (4.3). This makes regularization attractive for applications where obtaining a separate validation set is difficult or costly (where labelled data is scarce).

Instead of probing a hypothesis $\hat{h}$ on a validation set, regularization techniques compute estimate the loss increase when applying $\hat{h}$ to data points outside the training set. The loss increase is estimated by adding a regularization term to the training error in ERM (4.3).

© The Author(s), under exclusive license to Springer Nature Singapore Pte Ltd. 2022
A. Jung, *Machine Learning*, Machine Learning: Foundations, Methodologies, and Applications, https://doi.org/10.1007/978-981-16-8193-6_7

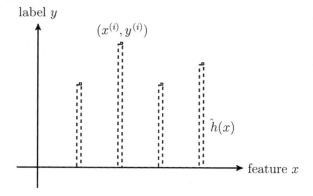

Fig. 7.1 The non-linear hypothesis map $\hat{h}$ perfectly fits the training set and has vanishing training error. Despite perfectly fitting the training set, the hypothesis $\hat{h}$ delivers the trivial (and useless) prediction $\hat{y} = \hat{h}(x) = 0$ for any datapoint that is not in the vicinity of the data points in the training set

Section 7.1 discusses the resulting regularized ERM, which we will refer to as structural risk minimization (SRM). It turns out that the SRM is equivalent to ERM using a smaller (pruned) hypothesis space. The amount of pruning depends on the weight of the regularization term relative to the training error. For an increasing weight of the regularization term, we obtain a stronger pruning resulting in a smaller effective hypothesis space.

Section 7.2 constructs regularization terms by requiring the resulting ML method to be robust against (small) random perturbations of the data points in a training set. Here, we replace each data point of a training set by the realization of a RV that fluctuates around this data point. This construction allows to interpret regularization as a (implicit) form of data augmentation.

Section 7.3 discusses data augmentation methods as a simulation-based implementation of regularization. Data augmentation adds a certain number of perturbed copies to each data point in the training set. One way to construct perturbed copies of a data point is to add (the realization of) a random vector to its features.

Section 7.4 analyzes the effect of regularization for linear regression using a simple probabilistic model for data points. This analysis parallels our previous study of the validation error of linear regression in Sect. 6.4. Similar to Sect. 6.4, we reveal a trade-off between the bias and variance of the hypothesis learnt by regularized linear regression. This trade- off was traced out by a discrete model parameter (the effective dimension) in Sect. 6.4. In contrast, regularization offers a continuous trade-off between bias and variance via a continuous regularization parameter.

Semi-supervised learning (SSL) uses (large amounts of) unlabeled data points to support the learning of a hypothesis from (a small number of) labeled data points [1]. Section 7.5 discusses SSL methods that use the statistical properties of unlabeled data points to construct useful regularization terms. These regularization terms are then used in SRM with a (typically small) set of labeled data points.

Multitask learning exploits similarities between different but related learning tasks [2]. We can formally define a learning task by a particular choice for the loss function (loss function) (see Sect. 2.3). The primary role of a loss function is to score the quality of a hypothesis map. However, the loss function also encapsulates the choice for the label of a data point. For learning tasks defined for a single underlying data generation process it is reasonable to assume that the same subset of features is relevant for those learning tasks. One example for such related learning tasks is a multi-label classification problem (see Section) where each individual label of a data point represents an separate learning task. Section 7.6 shows how multitask learning can be implemented using regularization methods. The loss incurred in different learning tasks serves mutual regularization terms in a joint SRM for all learning tasks.

Section 7.7 shows how regularization can be used for **transfer learning**. Like multitask learning also transfer learning exploits relations between different learning tasks. In contrast to multitask learning, which jointly solves the individual learning tasks, transfer learning solves the learning tasks sequentially. The most basic form of transfer learning is to fine tune a pre-trained model. A pre-trained model can be obtained via ERM (4.3) in a ("source") learning task for which we have a large amount of labeled training data. The fine-tuning is then obtained via ERM (4.3) in the ("target") learning task of interest for which we might have only a small amount of labeled training data.

7.1 Structural Risk Minimization

Section 2.2 defined the effective dimension d_{eff} ($\mathcal{H}$) of a hypothesis space $\mathcal{H}$ as the maximum number of data points that can be perfectly fit by some hypothesis $h \in \mathcal{H}$. As soon as the effective dimension of the hypothesis space in (4.3) exceeds the number m of training data points, we can find a hypothesis that perfectly fits the training data. However, a hypothesis that perfectly fits the training data might deliver poor predictions for data points outside the training set (see Fig. 7.1).

Modern ML methods typically use a hypothesis space with large effective dimension [3, 4]. Two well-known examples for such methods is linear regression (see Sect. 3.1) using a large number of features and deep learning with ANNs using a large number (billions) of artificial neurons (see Section 3.11). The effective dimension of these methods can be easily on the order of billions (10^9) if not larger [5]. To avoid overfitting during the naive use of ERM (4.3) we would require a training set containing at least as many data points as the effective dimension of the hypothesis space. However, in practice we often do not have access to training sets containing billions of labeled data points.

It seems natural to combat overfitting of a ML method by pruning its hypothesis space $\mathcal{H}$. We prune $\mathcal{H}$ by removing some of the hypothesis in $\mathcal{H}$ to obtain the smaller hypothesis space $\mathcal{H}' \subset \mathcal{H}$. We then replace ERM (4.3) with the restricted (or pruned) ERM

$$\hat{h} = \underset{h \in \mathcal{H}'}{\operatorname{argmin}} \widehat{L}(h|\mathcal{D}) \text{ with pruned hypothesis space } \mathcal{H}' \subset \mathcal{H}. \qquad (7.1)$$

The effective dimension of the pruned hypothesis space $\mathcal{H}'$ is typically much smaller than the effective dimension of the original (large) hypothesis space $\mathcal{H}$, $d_{\text{eff}}\left(\mathcal{H}'\right) \ll d_{\text{eff}}(\mathcal{H})$. For a given size m of the training set, the risk of overfitting in (7.1) is much smaller than the risk of overfitting in (4.3).

Example. Consider linear regression which the hypothesis space (3.1) constituted by linear maps $h(\mathbf{x}) = \mathbf{w}^T \mathbf{x}$. The effective dimension of (3.1) is equal to the number of features, $d_{\text{eff}}(\mathcal{H}) = n$. The hypothesis space $\mathcal{H}$ might be too large if we use a large number n of features, leading to overfitting. We prune (3.1) by retaining only linear hypotheses $h(\mathbf{x}) = \left(\mathbf{w}'\right)^T \mathbf{x}$ with weight vectors $\mathbf{w}'$ satisfying $w_3' = w_4' = \ldots = \mathbf{w}_n' = 0$. Thus, the hypothesis space $\mathcal{H}'$ is constituted by all linear maps that only depend on the first two features x_1, x_2 of a data point. The effective dimension of $\mathcal{H}'$ is dimension is $d_{\text{eff}}\left(\mathcal{H}'\right) = 2$ instead of $d_{\text{eff}}(\mathcal{H}) = n$.

Pruning the hypothesis space is a special case of a more general strategy which we refer to as SRM [6]. The idea behind SRM is to modify the training error in ERM (4.3) to favour hypotheses which are more smooth or regular in a specific sense. By enforcing a smooth hypothesis, a ML methods becomes less sensitive, or more robust, to small perturbations of the training data points. Section 7.2 discusses the intimate relation between the robustness (against perturbations of the training set) of a ML method and its ability to generalize to data points outside the training set.

We measure the smoothness of a hypothesis using a regularizer $\mathcal{R}(h) \in \mathbb{R}_+$. Roughly speaking, the value $\mathcal{R}(h)$ measures the irregularity or variation of a predictor map h. The (design) choice for the regularizer depends on the precise definition of what is meant by regularity or variation of a hypothesis. Section 7.3 discusses how a particular choice for the regularizer $\mathcal{R}(h)$ arises naturally from a probabilistic model for data points.

We obtain SRM by adding the scaled regularizer $\lambda \mathcal{R}(h)$ to the ERM (4.3),

$$\hat{h} = \underset{h \in \mathcal{H}}{\operatorname{argmin}} \left[\widehat{L}(h|\mathcal{D}) + \lambda \mathcal{R}(h)\right]$$

$$\overset{(2.16)}{=} \underset{h \in \mathcal{H}}{\operatorname{argmin}} \left[(1/m) \sum_{i=1}^{m} L((\mathbf{x}^{(i)}, y^{(i)}), h) + \lambda \mathcal{R}(h)\right]. \tag{7.2}$$

We can interpret the penalty term $\lambda \mathcal{R}(h)$ in (7.2) as an estimate (or approximation) for the increase, relative to the training error on $\mathcal{D}$, of the average loss of a hypothesis $\hat{h}$ when it is applied to data points outside $\mathcal{D}$. Another interpretation of the term $\lambda \mathcal{R}(h)$ will be discussed in Sect. 7.3.

The regularization parameter λ allows us to trade between a small training error $\widehat{L}(h^{(\mathbf{w})}|\mathcal{D})$ and small regularization term $\mathcal{R}(h)$, which enforces smoothness or regularity of h. If we choose a large value for λ, irregular or hypotheses h, with large $\mathcal{R}(h)$, are heavily "punished" in (7.2). Thus, increasing the value of λ results in the solution (minimizer) of (7.2) having smaller $\mathcal{R}(h)$. On the other hand, choosing a small value for λ in (7.2) puts more emphasis on obtaining a hypothesis h incurring a small training error. For the extreme case $\lambda = 0$, the SRM (7.2) reduces to ERM (4.3).

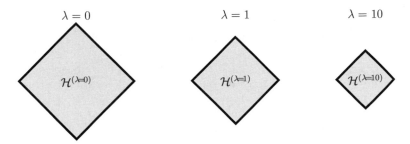

Fig. 7.2 Adding the scaled regularizer $\lambda \mathcal{R}(h)$ to the training error in the objective function of SRM (7.2) is equivalent to solving ERM (7.1) with a pruned hypothesis space $\mathcal{H}^{(\lambda)}$

The pruning approach (7.1) is intimately related to the SRM (7.2). They are, in a certain sense, **dual** to each other. First, note that (7.2) reduces to the pruning approach (7.1) when using the regularizer $\mathcal{R}(h) = 0$ for all $h \in \mathcal{H}'$, and $\mathcal{R}(h) = \infty$ otherwise, in (7.2). In the other direction, for many important choices for the regularizer $\mathcal{R}(h)$, there is a restriction $\mathcal{H}^{(\lambda)} \subset \mathcal{H}$ such that the solutions of (7.1) and (7.2) coincide (see Fig. 7.2). The relation between the optimization problems (7.1) and (7.2) can be made precise using the theory of convex duality (see [7, Ch. 5] and [8]).

For a hypothesis space $\mathcal{H}$ whose elements $h \in \mathcal{H}$ are parameterized by a weight vector $\mathbf{w} \in \mathbb{R}^n$, we can rewrite SRM (7.2) as

$$
\begin{aligned}
\widehat{\mathbf{w}}^{(\lambda)} &= \underset{\mathbf{w} \in \mathbb{R}^n}{\operatorname{argmin}} \left[\widehat{L}(h^{(\mathbf{w})} | \mathcal{D}) + \lambda \mathcal{R}(\mathbf{w}) \right] \\
&= \underset{\mathbf{w} \in \mathbb{R}^n}{\operatorname{argmin}} \left[(1/m) \sum_{i=1}^{m} L((\mathbf{x}^{(i)}, y^{(i)}), h^{(\mathbf{w})}) + \lambda \mathcal{R}(\mathbf{w}) \right].
\end{aligned} \tag{7.3}
$$

For the particular choice of squared error loss (2.8), linear hypothesis space (3.1) and regularizer $\mathcal{R}(\mathbf{w}) = \|\mathbf{w}\|_2^2$, SRM (7.3) specializes to

$$
\widehat{\mathbf{w}}^{(\lambda)} = \underset{\mathbf{w} \in \mathbb{R}^n}{\operatorname{argmin}} \left[(1/m) \sum_{i=1}^{m} \left(y^{(i)} - \mathbf{w}^T \mathbf{x}^{(i)} \right)^2 + \lambda \|\mathbf{w}\|_2^2 \right]. \tag{7.4}
$$

The special case (7.4) of SRM (7.3) is known as ridge regression [9].

Ridge regression (7.4) is equivalent to (see [8, Ch. 5])

$$
\widehat{\mathbf{w}}^{(\lambda)} = \underset{h^{(\mathbf{w})} \in \mathcal{H}^{(\lambda)}}{\operatorname{argmin}} (1/m) \sum_{i=1}^{m} \left(y^{(i)} - h^{(\mathbf{w})}(\mathbf{x}^{(i)}) \right)^2 \tag{7.5}
$$

with the restricted hypothesis space

$$\mathcal{H}^{(\lambda)} := \{h^{(\mathbf{w})} : \mathbb{R}^n \to \mathbb{R} : h^{(\mathbf{w})}(\mathbf{x}) = \mathbf{w}^T \mathbf{x},$$

with some weights (weights) $\mathbf{w}$ satisfying $\|\mathbf{w}\|_2^2 \le C(\lambda)\} \subset \mathcal{H}^{(n)}.$ (7.6)

For any given value λ of the regularization parameter in (7.4), there is a number $C(\lambda)$ such that solutions of (7.4) coincide with the solutions of (7.5). Thus, ridge regression (7.4) i is equivalent to linear regression using a pruned version $\mathcal{H}^{(\lambda)}$ of the linear hypothesis space (3.1). The pruned hypothesis space $\mathcal{H}^{(\lambda)}$ (7.6) depends varies continuously with the regularization parameter λ.

Another popular special case of ERM (7.3) is obtained for the regularizer $\mathcal{R}(\mathbf{w}) = \|\mathbf{w}\|_1$ and known as the Lasso [10]

$$\widehat{\mathbf{w}}^{(\lambda)} = \operatorname*{argmin}_{\mathbf{w} \in \mathbb{R}^n} \Big[(1/m) \sum_{i=1}^m \big(y^{(i)} - \mathbf{w}^T \mathbf{x}^{(i)} \big)^2 + \lambda \|\mathbf{w}\|_1 \Big]. \tag{7.7}$$

Ridge regression (7.4) and the Lasso (7.7) have fundamentally different computational and statistical properties. Involving a smooth and convex objective function, ridge regression (7.4) can be implemented using efficient GD methods. The objective function of Lasso (7.7) is also convex but non-smooth and therefore requires advanced optimization methods. The increased computational complexity of Lasso (7.7) comes at the benefit of typically delivering a hypothesis with a smaller risk than those obtained from ridge regression [4, 10].

7.2   Robustness

Section 7.1 motivates regularization as a soft variant of model selection. Indeed, the regularization term in SRM (7.2) is equivalent to ERM (7.1) using a pruned (reducing) hypothesis space. We now discuss an alternative view on regularization as a means to make ML methods robust.

The ML methods discussed in Chap. 4 rest on the idealizing assumption that we have access to the true label values and feature values of labeled data points (the training set). These methods learn a hypothesis $h \in \mathcal{H}$ with minimum average loss (training error) incurred for data points in the training set. In practice, the acquisition of label and feature values might be prone to errors. These errors might stem from the measurement device itself (hardware failures or thermal noise) or might be due to human mistakes such as labelling errors.

Let us assume for the sake of exposition that the label values $y^{(i)}$ in the training set are accurate but that the features $\mathbf{x}^{(i)}$ are a perturbed version of the true features of the ith data point. Thus, instead of having observed the data point $(\mathbf{x}^{(i)}, y^{(i)})$ we could have equally well observed the data point $(\mathbf{x}^{(i)} + \boldsymbol{\varepsilon}, y^{(i)})$ in the training set. Here, we have modelled the perturbations in the features using a RV $\boldsymbol{\varepsilon}$. The probability distribution of the perturbation $\boldsymbol{\varepsilon}$ is a design parameter that controls robustness properties of the overall ML method. We will study a particular choice for this distribution in Sect. 7.3.

A robust ML method should learn a hypothesis that incurs a small loss not only for a specific data point $(\mathbf{x}^{(i)}, y^{(i)})$ but also for perturbed data points $(\mathbf{x}^{(i)} + \boldsymbol{\varepsilon}, y^{(i)})$. Therefore, it seems natural to replace the loss $L((\mathbf{x}^{(i)}, y^{(i)}), h)$, incurred on the ith data point in the training set, with the expectation

$$\mathbb{E}\{L((\mathbf{x}^{(i)} + \boldsymbol{\varepsilon}, y^{(i)}), h)\}. \tag{7.8}$$

The expectation (7.8) is computed using the probability distribution of the perturbation $\boldsymbol{\varepsilon}$. We will show in Sect. 7.3 that minimizing the average of the expectation (7.8), for $i = 1, \ldots, m$, is equivalent to the SRM (7.2).

Using the expected loss (7.8) is not the only possible approach to make a ML method robust. Another approach to make a ML method robust is known as bagging. The idea of bagging is to use the bootstrap method (see Sect. 6.5 and [9, Chap. 8]) to construct a finite number of perturbed copies $\mathcal{D}^{(1)}, \ldots, \mathcal{D}^{(B)}$ of the original training set $\mathcal{D}$.

We then learn (e.g, using ERM) a separate hypothesis $h^{(b)}$ for each perturbed copy $\mathcal{D}^{(b)}$, $b = 1, \ldots, B$. This results in a whole ensemble of different hypotheses $h^{(b)}$ which might even belong to different hypothesis spaces. For example, one the hypotheis $h^{(1)}$ could be a linear map (see Sect. 3.1) and the hypothesis $h^{(2)}$ could be obtained from an ANN (see Sect. 3.11).

The final hypothesis delivered by bagging is obtained by combining or aggregating (e.g., using the average) the predictions $h^{(b)}(\mathbf{x})$ delivered by each hypothesis $h^{(b)}$, for $b = 1, \ldots, B$ in the ensemble. The ML method referred to as random forest uses bagging to learn an ensemble of decision trees (see Sect. 3.10). The individual predictions obtained from the trees in a random forest are combined (e.g., using an average in regression or a majority vote in binary classification) to obtain a final prediction [9].

7.3 Data Augmentation

ML methods using ERM (4.3) are prone to overfitting as soon as the effective dimension of the hypothesis space $\mathcal{H}$ exceeds the number m of training data points. Sections 6.3 and 7.1 approached this by modifying either the model or the loss function by adding a regularization term. Both approaches prune the hypothesis space $\mathcal{H}$ underlying a ML method to reduce the effective dimension $d_{\text{eff}}(\mathcal{H})$. Model selection does this reduction in a discrete fashion while regularization implements a soft "shrinking" of the hypothesis space.

Instead of trying to reduce the effective dimension we could also try to increase the number m of training data points used in ERM (4.3). We now discuss how to synthetically generate new labeled data points by exploiting known structures that are inherent to a given application domain.

The data arising in many ML applications exhibit intrinsic symmetries and invariances at least in some approximation. The rotated image of a cat still shows a cat. The temperature measurement taken at a given location will be similar to another measurement taken 10 milliseconds later. Data augmentation exploits such symmetries and invariances to augment the raw data with additional synthetic data.

Let us illustrate data augmentation using an application that involves data points characterized by features $\mathbf{x} \in \mathbb{R}^n$ and number labels $y \in \mathbb{R}$. We assume that the data generating process is such that data points with close feature values have the same label. Equivalently, this assumption is requiring the resulting ML method to be robust against small perturbations of the feature values (see Sect. 7.2). This suggests to augment a data point $(\mathbf{x}, y)$ by several synthetic data points

$$(\mathbf{x} + \boldsymbol{\varepsilon}^{(1)}, y), \ldots, (\mathbf{x} + \boldsymbol{\varepsilon}^{(B)}, y), \tag{7.9}$$

with $\boldsymbol{\varepsilon}^{(1)}, \ldots, \boldsymbol{\varepsilon}^{(B)}$ being realizations of independent and identically distributed (i.i.d.) (i.i.d.) random vectors with the same probability distribution $p(\boldsymbol{\varepsilon})$.

Given a (raw) dataset $\mathcal{D} = \{(\mathbf{x}^{(1)}, y^{(1)}), \ldots, (\mathbf{x}^{(m)}, y^{(m)})\}$ we denote the associated augmented dataset by

$$\begin{aligned} \mathcal{D}' = \{ & (\mathbf{x}^{(1,1)}, y^{(1)}), \ldots, (\mathbf{x}^{(1,B)}, y^{(1)}), \\ & (\mathbf{x}^{(2,1)}, y^{(2)}), \ldots, (\mathbf{x}^{(2,B)}, y^{(2)}), \\ & \ldots \\ & (\mathbf{x}^{(m,1)}, y^{(m)}), \ldots, (\mathbf{x}^{(m,B)}, y^{(m)})\}. \end{aligned} \tag{7.10}$$

The size of the augmented dataset $\mathcal{D}'$ is $m' = B \times m$. For a sufficiently large augmentation parameter B, the augmented sample size m' is larger than the effective dimension n of the hypothesis space $\mathcal{H}$. We then learn a hypothesis via ERM on the augmented dataset,

$$\begin{aligned} \hat{h} & = \underset{h \in \mathcal{H}}{\operatorname{argmin}} \, \widehat{L}(h|\mathcal{D}') \\ & \overset{(7.10)}{=} \underset{h \in \mathcal{H}}{\operatorname{argmin}} (1/m') \sum_{i=1}^{m} \sum_{b=1}^{B} L((\mathbf{x}^{(i,b)}, y^{(i,b)}), h) \\ & \overset{(7.9)}{=} \underset{h \in \mathcal{H}}{\operatorname{argmin}} (1/m) \sum_{i=1}^{m} (1/B) \sum_{b=1}^{B} L((\mathbf{x}^{(i)} + \boldsymbol{\varepsilon}^{(b)}, y^{(i)}), h). \end{aligned} \tag{7.11}$$

We can interpret data-augmented ERM (7.11) as a data-driven form of regularization (see Sect. 7.1). The regularization is implemented by replacing, for each data point $(\mathbf{x}^{(i)}, y^{(i)}) \in \mathcal{D}$, the loss $L((\mathbf{x}^{(i)}, y^{(i)}), h)$ with the average loss $(1/B) \sum_{b=1}^{B} L((\mathbf{x}^{(i)} + \boldsymbol{\varepsilon}^{(b)}, y^{(i)}), h)$ over the augmented data points that accompany $(\mathbf{x}^{(i)}, y^{(i)}) \in \mathcal{D}$.

Note that in order to implement (7.11) we need to first generate B realizations $\boldsymbol{\varepsilon}^{(b)} \in \mathbb{R}^n$ of i.i.d. random vectors with common probability distribution $p(\boldsymbol{\varepsilon})$. This might be computationally costly for a large B, n. However, when using a large augmentation parameter B, we might use the approximation

$$(1/B) \sum_{b=1}^{B} L((\mathbf{x}^{(i)} + \boldsymbol{\varepsilon}^{(b)}, y^{(i)}), h) \approx \mathbb{E}\{L((\mathbf{x}^{(i)} + \boldsymbol{\varepsilon}, y^{(i)}), h)\}. \tag{7.12}$$

This approximation is made precise by a key result of probability theory, known as the law of large numbers. We obtain an instance of ERM by inserting (7.12) into (7.11),

$$\hat{h} = \underset{h \in \mathcal{H}}{\operatorname{argmin}} (1/m) \sum_{i=1}^{m} \mathbb{E}\{L((\mathbf{x}^{(i)} + \boldsymbol{\varepsilon}, y^{(i)}), h)\}. \tag{7.13}$$

The usefulness of (7.13) as an approximation to the augmented ERM (7.11) depends on the difficulty of computing the expectation $\mathbb{E}\{L((\mathbf{x}^{(i)} + \boldsymbol{\varepsilon}, y^{(i)}), h)\}$. The complexity of computing this expectation depends on the choice of loss function and the choice for the probability distribution $p(\boldsymbol{\varepsilon})$.

Let us study (7.13) for the special case linear regression with squared error loss (2.8) and linear hypothesis space (3.1),

$$\hat{h} = \underset{h^{(\mathbf{w})} \in \mathcal{H}^{(n)}}{\operatorname{argmin}} (1/m) \sum_{i=1}^{m} \mathbb{E}\{(y^{(i)} - \mathbf{w}^T(\mathbf{x}^{(i)} + \boldsymbol{\varepsilon}))^2\}. \tag{7.14}$$

We use perturbations $\boldsymbol{\varepsilon}$ drawn a multivariate normal distribution with zero mean and covariance matrix $\sigma^2\mathbf{I}$,

$$\boldsymbol{\varepsilon} \sim \mathcal{N}(\mathbf{0}, \sigma^2\mathbf{I}). \tag{7.15}$$

We develop (7.14) further by using

$$\mathbb{E}\{(y^{(i)} - \mathbf{w}^T\mathbf{x}^{(i)})\boldsymbol{\varepsilon}\} = \mathbf{0}. \tag{7.16}$$

The identity (7.16) uses that the data points $(\mathbf{x}^{(i)}, y^{(i)})$ are fixed and known (deterministic) while $\boldsymbol{\varepsilon}$ is a zero-mean random vector. Combining (7.16) with (7.14),

$$\mathbb{E}\{(y^{(i)} - \mathbf{w}^T(\mathbf{x}^{(i)} + \boldsymbol{\varepsilon}))^2\} = (y^{(i)} - \mathbf{w}^T\mathbf{x}^{(i)})^2 + \|\mathbf{w}\|_2^2 \mathbb{E}\{\|\boldsymbol{\varepsilon}\|_2^2\}$$

$$= (y^{(i)} - \mathbf{w}^T\mathbf{x}^{(i)})^2 + n\|\mathbf{w}\|^2\sigma^2. \tag{7.17}$$

where the last step used $\mathbb{E}\{\|\boldsymbol{\varepsilon}\|_2^2\} \overset{(7.15)}{=} n\sigma^2$. Inserting (7.17) into (7.14),

$$\hat{h} = \underset{h^{(\mathbf{w})} \in \mathcal{H}^{(n)}}{\text{argmin}} (1/m) \sum_{i=1}^{m} \left(y^{(i)} - \mathbf{w}^T \mathbf{x}^{(i)}\right)^2 + n \|\mathbf{w}\|^2 \sigma^2. \tag{7.18}$$

We have obtained (7.18) as an approximation of the augmented ERM (7.11) for the special case of squared error loss (2.8) and the linear hypothesis space (3.1). This approximation uses the law of large numbers (7.12) and becomes more accurate for increasing augmentation parameter B.

Note that (7.18) is nothing but ridge regression (7.4) using the regularization parameter $\lambda = n\sigma^2$. Thus, we can interpret ridge regression as implicit data augmentation (7.10) by applying random perturbations (7.9) to the feature vectors in the original training set $\mathcal{D}$.

The regularizer $\mathcal{R}(\mathbf{w}) = \|\mathbf{w}\|_2^2$ in (7.18) arose naturally from the specific choice for the probability distribution (7.15) of the random perturbation $\boldsymbol{\varepsilon}^{(i)}$ in (7.9) and using the squared error loss. Other choices for this probability distribution or the loss function result in different regularizers.

Augmenting data points with random perturbations distributed according (7.15) treat the features of a data point independently. For application domains that generate data points with highly correlated features it might be useful to augment data points using random perturbations $\boldsymbol{\varepsilon}$ (see (7.9)) distributed as

$$\boldsymbol{\varepsilon} \sim \mathcal{N}(\mathbf{0}, \mathbf{C}). \tag{7.19}$$

The covariance matrix $\mathbf{C}$ of the perturbation $\boldsymbol{\varepsilon}$ can be chosen using domain expertise or estimated (see Sect. 7.5). Inserting the distribution (7.19) into (7.13),

$$\hat{h} = \underset{h^{(\mathbf{w})} \in \mathcal{H}^{(n)}}{\text{argmin}} \left[(1/m) \sum_{i=1}^{m} \left(y^{(i)} - \mathbf{w}^T \mathbf{x}^{(i)}\right)^2 + \mathbf{w}^T \mathbf{C} \mathbf{w} \right]. \tag{7.20}$$

Note that (7.20) reduces to ordinary ridge regression (7.18) for the choice $\mathbf{C} = \sigma^2 \mathbf{I}$.

7.4   Statistical and Computational Aspects of Regularization

The goal of this section is to develop a better understanding for the effect of the regularization term in SRM (7.3). We will analyze the solutions of ridge regression (7.4) which is the special case of SRM using the linear hypothesis space (3.1) and squared error loss (2.8). Using the feature matrix $\mathbf{X} = \left(\mathbf{x}^{(1)}, \ldots, \mathbf{x}^{(m)}\right)^T$ and label vector $\mathbf{y} = (y^{(1)}, \ldots, y^{(m)})^T$, we can rewrite (7.4) more compactly as

$$\widehat{\mathbf{w}}^{(\lambda)} = \underset{\mathbf{w} \in \mathbb{R}^n}{\text{argmin}} \left[(1/m) \|\mathbf{y} - \mathbf{X}\mathbf{w}\|_2^2 + \lambda \|\mathbf{w}\|_2^2 \right]. \tag{7.21}$$

The solution of (7.21) is given by

$$\widehat{\mathbf{w}}^{(\lambda)} = (1/m)((1/m)\mathbf{X}^T\mathbf{X} + \lambda\mathbf{I})^{-1}\mathbf{X}^T\mathbf{y}. \tag{7.22}$$

For $\lambda = 0$, (7.22) reduces to the formula (6.17) for the optimal weights in linear regression (see (7.4) and (4.5)). Note that for $\lambda > 0$, the formula (7.22) is always valid, even when $\mathbf{X}^T\mathbf{X}$ is singular (not invertible). For $\lambda > 0$ the optimization problem (7.21) (and (7.4)) has the unique solution (7.22).

To study the statistical properties of the predictor $h^{(\widehat{\mathbf{w}}^{(\lambda)})}(\mathbf{x}) = (\widehat{\mathbf{w}}^{(\lambda)})^T\mathbf{x}$ (see (7.22)) we use the probabilistic toy model (6.13), (6.15) and (6.16) that we used already in Sect. 6.4. We interpret the training data $\mathcal{D}^{(\text{train})} = \{(\mathbf{x}^{(i)}, y^{(i)})\}_{i=1}^m$ as realizations of i.i.d. RVs whose distribution is defined by (6.13), (6.15) and (6.16).

We can then define the average prediction error of ridge regression as

$$E_{\text{pred}}^{(\lambda)} := \mathbb{E}\left\{\left(y - h^{(\widehat{\mathbf{w}}^{(\lambda)})}(\mathbf{x})\right)^2\right\}. \tag{7.23}$$

As shown in Sect. 6.4, the error $E_{\text{pred}}^{(\lambda)}$ is the sum of three components: the bias, the variance and the noise variance σ^2 (see (6.27)). The bias of $\widehat{\mathbf{w}}^{(\lambda)}$ is

$$B^2 = \left\|(\mathbf{I} - \mathbb{E}\{(\mathbf{X}^T\mathbf{X} + m\lambda\mathbf{I})^{-1}\mathbf{X}^T\mathbf{X}\})\overline{\mathbf{w}}\right\|_2^2. \tag{7.24}$$

For sufficiently large size m of the training set, we can use the approximation

$$\mathbf{X}^T\mathbf{X} \approx m\mathbf{I} \tag{7.25}$$

such that (7.24) can be approximated as

$$B^2 \approx \left\|(\mathbf{I} - (\mathbf{I} + \lambda\mathbf{I})^{-1})\overline{\mathbf{w}}\right\|_2^2$$
$$= \sum_{l=1}^n \frac{\lambda}{1+\lambda}\overline{w}_l^2. \tag{7.26}$$

Let us compare the (approximate) bias term (7.26) of regularized linear regression with the bias term (6.23) of ordinary linear regression (which is the extreme case of ridge regression with $\lambda = 0$). The bias term (7.26) increases with increasing regularization parameter λ in ridge regression (7.4). In many relevant settings, the increase in bias is outweighed by the reduction in variance. The variance typically decreases with increasing λ as shown next.

The variance of ridge regression (7.4) satisfies

$$V = (\sigma^2/m^2) \times$$
$$\text{tr}\{\mathbb{E}\{((1/m)\mathbf{X}^T\mathbf{X} + \lambda\mathbf{I})^{-1}\mathbf{X}^T\mathbf{X}((1/m)\mathbf{X}^T\mathbf{X} + \lambda\mathbf{I})^{-1}\}\}. \tag{7.27}$$

Fig. 7.3 The bias and
variance of regularized linear
regression depend on the
regularization parameter λ in
an opposite manner resulting
in a bias-variance trade-off

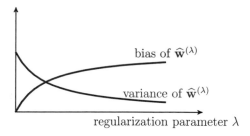

Inserting the approximation (7.25) into (7.27),

$$V \approx \sigma^2 (1/m)(n/(1+\lambda)). \tag{7.28}$$

According to (7.28), the variance of $\widehat{\mathbf{w}}^{(\lambda)}$ decreases with increasing regularization parameter λ of ridge regression (7.4). This is the opposite behaviour as observed for the bias (7.26), which increases with increasing λ. The approximate variance formula (7.28) suggests to interpret the ratio $(n/(1+\lambda))$ as the effective number of features used by ridge regression. Increasing the regularization parameter λ decreases the effective number of features.

Figure 7.3 illustrates the trade-off between the bias B^2 (7.26) of ridge regression, which increases for increasing λ, and the variance V (7.28) which decreases with increasing λ. Note that we have seen another example for a bias-variance trade-off in Sect. 6.4. This trade-off was traced out by a discrete (model complexity) parameter $r \in \{1, 2, \ldots\}$ (see (6.14)). In stark contrast to discrete model selection, the bias-variance trade-off for ridge regression is traced out by the continuous regularization parameter $\lambda \in \mathbb{R}_+$.

The main statistical effect of the regularization term in ridge regression is to balance the bias with the variance to minimize the average prediction error of the learnt hypothesis. There is also a computational effect or adding a regularization term. Roughly speaking, the regularization term serves as a pre-conditioning of the optimization problem and, in turn, reduces the computational complexity of solving ridge regression (7.21).

The objective function in (7.21) is a smooth (infinitely often differentiable) convex function. We can therefore use GD to solve (7.21) efficiently (see Chap. 5). Algorithm 8 summarizes the application of GD to (7.21). The computational complexity of Algorithm 8 depends crucially on the number of GD iterations required to reach a sufficiently small neighbourhood of the solutions to (7.21). Adding the regularization term $\lambda \|\mathbf{w}\|_2^2$ to the objective function of linear regression **speeds up GD**. To verify this claim, we first rewrite (7.21) as the quadratic problem

$$\min_{\mathbf{w}\in\mathbb{R}^n} \underbrace{(1/2)\mathbf{w}^T\mathbf{Q}\mathbf{w} - \mathbf{q}^T\mathbf{w}}_{=f(\mathbf{w})}$$

$$\text{with } \mathbf{Q} = (1/m)\mathbf{X}^T\mathbf{X} + \lambda\mathbf{I}, \mathbf{q} = (1/m)\mathbf{X}^T\mathbf{y}. \tag{7.29}$$

This is similar to the quadratic optimization problem (4.9) underlying linear regression but with a different matrix $\mathbf{Q}$. The computational complexity (number of iterations) required by GD (see (5.4)) applied to solve (7.29) up to a prescribed accuracy depends crucially on the condition number $\kappa(\mathbf{Q}) \geq 1$ of the psd matrix $\mathbf{Q}$ [11]. The smaller the condition number $\kappa(\mathbf{Q})$, the fewer iterations are required by GD. A matrix with small condition number is also referred to as being "well-conditioned".

The condition number of the matrix $\mathbf{Q}$ in (7.29) is given by

$$\kappa(\mathbf{Q}) = \frac{\lambda_{\max}((1/m)\mathbf{X}^T\mathbf{X}) + \lambda}{\lambda_{\min}((1/m)\mathbf{X}^T\mathbf{X}) + \lambda}. \tag{7.30}$$

According to (7.30), the condition number tends to one for increasing regularization parameter λ,

$$\lim_{\lambda\to\infty} \frac{\lambda_{\max}((1/m)\mathbf{X}^T\mathbf{X}) + \lambda}{\lambda_{\min}((1/m)\mathbf{X}^T\mathbf{X}) + \lambda} = 1. \tag{7.31}$$

Thus, the number of required GD iterations in Algorithm 8 decreases with increasing regularization parameter λ.

Algorithm 8 Regularized Linear regression via GD

Input: dataset $\mathcal{D} = \{(\mathbf{x}^{(i)}, y^{(i)})\}_{i=1}^m$; GD step size $\alpha > 0$.
Initialize: set $\mathbf{w}^{(0)} := \mathbf{0}$; set iteration counter $k := 0$
1: **repeat**
2:     $r := r + 1$ (increase iteration counter)
3:     $\mathbf{w}^{(r)} := (1 - \alpha\lambda)\mathbf{w}^{(r-1)} + \alpha(2/m)\sum_{i=1}^m (y^{(i)} - (\mathbf{w}^{(r-1)})^T\mathbf{x}^{(i)})\mathbf{x}^{(i)}$ (do a GD step (5.4))
4: **until** stopping criterion met
Output: $\mathbf{w}^{(r)}$ (which approximates $\widehat{\mathbf{w}}^{(\lambda)}$ in (7.21))

7.5 Semi-Supervised Learning

Consider the task of predicting the numeric label y of a data point $\mathbf{z} = (\mathbf{x}, y)$ based on its feature vector $\mathbf{x} = (x_1, \ldots, x_n)^T \in \mathbb{R}^n$. At our disposal are two datasets $\mathcal{D}^{(u)}$ and $\mathcal{D}^{(l)}$. For each data point in $\mathcal{D}^{(u)}$ we only know the feature vector. We therefore refer to $\mathcal{D}^{(u)}$ as "unlabelled data". For each data point in $\mathcal{D}^{(l)}$ we know both, the feature vector $\mathbf{x}$ and the label y. We therefore refer to $\mathcal{D}^{(l)}$ as "labeled data".

SSL methods exploit the information provided by unlabelled data $\mathcal{D}^{(u)}$ to support the learning of a hypothesis based on minimizing its empirical risk on the labelled (training) data $\mathcal{D}^{(l)}$. The success of SSL methods depends on the statistical properties of the data generated within a given application domain. Loosely speaking, the information provided by the probability distribution of the features must be relevant for the ultimate task of predicting the label y from the features $\mathbf{x}$ [1].

Let us design a SSL method, summarized in Algorithm 9 below, using the data augmentation perspective from Sect. 7.3. The idea is the augment the (small) labeled dataset $\mathcal{D}^{(l)}$ by adding random perturbations for the features vectors of data point in $\mathcal{D}^{(l)}$. This is reasonable for applications where feature vectors are subject to inherent measurement or modelling errors. Given a data point with vector $\mathbf{x}$ we could have equally well observed a feature vector $\mathbf{x} + \boldsymbol{\varepsilon}$ with some small random perturbation $\boldsymbol{\varepsilon} \sim \mathcal{N}(\mathbf{0}, \mathbf{C})$. To estimate the covariance matrix $\mathbf{C}$, we use the sample covariance matrix of the feature vectors in the (large) unlabelled dataset $\mathcal{D}^{(u)}$. We then learn a hypothesis using the augmented (regularized) ERM (7.20).

Algorithm 9 A Semi-Supervised Learning Algorithm

Input: labeled dataset $\mathcal{D}^{(l)} = \{(\mathbf{x}^{(i)}, y^{(i)})\}_{i=1}^{m}$; unlabeled dataset $\mathcal{D}^{(u)} = \{\widetilde{\mathbf{x}}^{(i)}\}_{i=1}^{m'}$
1: compute $\mathbf{C}$ via sample covariance on $\mathcal{D}^{(u)}$,

$$\mathbf{C} := (1/m') \sum_{i=1}^{m'} \left(\widetilde{\mathbf{x}}^{(i)} - \widehat{\mathbf{x}}\right)\left(\widetilde{\mathbf{x}}^{(i)} - \widehat{\mathbf{x}}\right)^{T} \text{ with } \widehat{\mathbf{x}} := (1/m') \sum_{i=1}^{m'} \widetilde{\mathbf{x}}^{(i)}. \tag{7.32}$$

2: compute (e.g. using GD steps (5.4))

$$\widehat{\mathbf{w}} := \operatorname*{argmin}_{\mathbf{w} \in \mathbb{R}^n} \left[(1/m) \sum_{i=1}^{m} \left(y^{(i)} - \mathbf{w}^T \mathbf{x}^{(i)}\right)^2 + \mathbf{w}^T \mathbf{C} \mathbf{w} \right]. \tag{7.33}$$

Output: hypothesis $\widehat{h}(\mathbf{x}) = (\widehat{\mathbf{w}})^T \mathbf{x}$

7.6 Multitask Learning

We can identify a learning task with the loss function $L((\mathbf{x}, y), h)$ that is used to measure the quality of a particular hypothesis $h \in \mathcal{H}$. Note that the loss obtained for a given data point also depends on the definition for the label of a data point. For the same data points, we obtain different learning tasks from different choices or definitions for the label of a data point. Multitask learning exploits the similarities between different learning tasks to jointly solve them.

Example. Consider a data point $\mathbf{z}$ representing a hand-drawing that is collected via the online game https://quickdraw.withgoogle.com/. The features of a data point are the pixel intensities of the bitmap which is used to store the hand-drawing. As label we could use the fact if a hand-drawing shows an apple or not. This results in the learning task $\mathcal{T}^{(1)}$. Another choice for the label of a hand-drawing could be the fact if a hand-drawing shows a fruit at all or not. This results in another learning task $\mathcal{T}^{(2)}$ which is similar but different from the task $\mathcal{T}^{(1)}$.

The idea of multitask learning is that a reasonable hypothesis h for a learning task should also do well for a related learning tasks. Thus, we can use the loss incurred on similar learning tasks as a regularization term for learning a hypothesis for the learning task at hand. Algorithm 10 is a straightforward implementation of this idea for a given dataset that gives rise to T related learning tasks $\mathcal{T}^{(1)}, \ldots, \mathcal{T}^{(T)}$. For each individual learning task $\mathcal{T}^{(t')}$ it uses the loss on the remaining learning tasks $\mathcal{T}^{(t)}$, with $t \neq t'$, as regularization term in (7.34).

Algorithm 10 A Multitask Learning Algorithm

Input: dataset $\mathcal{D} = \{\mathbf{z}^{(1)}, \ldots, \mathbf{z}^{(m)}\}$ with T associated learning tasks with loss functions $L^{(1)}, \ldots, L^{(T)}$, hypothesis space $\mathcal{H}$
1: learn a hypothesis $\hat{h}$ via

$$\hat{h} := \underset{h \in \mathcal{H}}{\operatorname{argmin}} \sum_{t=1}^{T} \sum_{i=1}^{m} L^{(t)}(\mathbf{z}^{(i)}, h). \tag{7.34}$$

Output: hypothesis $\hat{h}$

The applicability of Algorithm 10 is somewhat limited as it aims at finding a single hypothesis that does well for all T learning tasks simultaneously. For certain application domains it might be more reasonable to not learn a single hypothesis for all learning tasks but to learn a separate hypothesis $h^{(t)}$ for each learning task $t = 1, \ldots, T$. However, these separate hypotheses typically might still share some structural similarities.[1] We can enforce different notion of similarities between the hypotheses $h^{(t)}$ by adding a regularization term to the loss functions of the tasks.

Algorithm 11 generalizes Algorithms 10 by learning a separate hypothesis for each task t while requiring these hypotheses to be structurally similar. The structural (dis-)similarity between the hypotheses is measured by a regularization term $\mathcal{R}$.

7.7 Transfer Learning

Regularization is also instrumental for transfer learning to capitalize on synergies between different related learning tasks [13, 14]. Transfer learning is enabled by constructing regularization terms for a learning task by using the result of a previous

---

[1] One important example for such a structural similarity in the case of linear predictors $h^{(t)}(\mathbf{x}) = \left(\mathbf{w}^{(t)}\right)^T \mathbf{x}$ is when the weight vectors $\mathbf{w}^{(T)}$ have a small joint support $\bigcup_{t=1,\ldots,T} \operatorname{supp}(w^{(t)})$. Requiring the weight vectors to have a small joint support is equivalent to requiring the stacked vector $\widetilde{\mathbf{w}} = \left(\mathbf{w}^{(1)}, \ldots, \mathbf{w}^{(T)}\right)$ to be block (group) sparse [12].

Algorithm 11 A Multitask Learning Algorithm

Input: dataset $\mathcal{D} = \{\mathbf{z}^{(1)}, \ldots, \mathbf{z}^{(m)}\}$ with T associated learning tasks with loss functions $L^{(1)}, \ldots, L^{(T)}$, hypothesis space $\mathcal{H}$

1: learn a hypothesis $\hat{h}$ via

$$\hat{h}^{(1)}, \ldots, \hat{h}^{(T)} := \underset{h^{(1)}, \ldots, h^{(T)} \in \mathcal{H}}{\mathrm{argmin}} \sum_{t=1}^{T} \sum_{i=1}^{m} L^{(t)}\big(\mathbf{z}^{(i)}, h^{(t)}\big) + \lambda \mathcal{R}\big(h^{(1)}, \ldots, h^{(T)}\big). \qquad (7.35)$$

Output: hypotheses $\hat{h}^{(1)}, \ldots, \hat{h}^{(T)}$

---

leaning task. While multitask learning methods solve many related learning tasks simultaneously, transfer learning methods operate in a sequential fashion.

To illustrate the idea of transfer learning consider two learning tasks which differ in their intrinsic difficulty. We consider a learning task to be easy if it involves if we can easily gather large amounts of labeled (training) data for that task. Consider the learning task $\mathcal{T}^{(1)}$ of predicting whether an image shows a cat or not. For this learning task we can easily gather a large training set $\mathcal{D}^{(1)}$ using via image collections of animals. Another (related) learning task $\mathcal{T}^{(2)}$ is to predicting whether an image shows a cat of a particular breed, with a particular body height and with a specific age, we might not be able to collect many labeled data points.

7.8 Exercises

Exercise 7.1 Ridge Regression is a Quadratic Problem. Consider the linear hypothesis space consisting of linear maps parameterized by weights $\mathbf{w}$. We try to find the best linear map by minimizing the regularized average squared error loss (empirical risk) incurred on some labeled data points $(\mathbf{x}^{(1)}, y^{(1)}), (\mathbf{x}^{(2)}, y^{(2)}), \ldots, (\mathbf{x}^{(m)}, y^{(m)})$. As the regularizer we use $\|\mathbf{w}\|^2$, yielding the following learning problem

$$\min_{\mathbf{w} \in \mathbb{R}^n} f(\mathbf{w}) = (1/m) \sum_{i=1}^{m} \big(y^{(i)} - \mathbf{w}^T \mathbf{x}^{(i)}\big) + \|\mathbf{w}\|_2^2.$$

Is it possible to rewrite the objective function $f(\mathbf{w})$ as a convex quadratic function $f(\mathbf{w}) = \mathbf{w}^T \mathbf{C} \mathbf{w} + \mathbf{b}\mathbf{w} + c$? If this is possible, how are the matrix $\mathbf{C}$, vector $\mathbf{b}$ and constant c related to the feature vectors and labels of the training data?

Exercise 7.2 Regularization or Model Selection. Consider data points, each characterized by $n = 10$ features $\mathbf{x} \in \mathbb{R}^n$ and a single numeric label y. We want to learn a linear hypothesis $h(\mathbf{x}) = \mathbf{w}^T \mathbf{x}$ by minimizing the average squared error on the training set $\mathcal{D}$ of size $m = 4$. We could learn such a hypothesis by two approaches. The first approach is to split the dataset into a training set and a validation set. Then we consider all models that consists of linear hypotheses with weight vectors having at most two non-zero weights. Each of these models corresponds to a different subset of two weights that might be non-zero. Find the model resulting in the smallest validation errors (see Algorithm 5). Compute the average loss of the resulting optimal linear hypothesis on some data points that have neither been used for training nor for validation. Compare this average loss ("test error") with the average loss obtained on the same data points by the hypothesis learnt by ridge regression (7.4).

References

1. O. Chapelle, B. Schölkopf, A. Zien (eds.), *Semi-Supervised Learning* (The MIT Press, Cambridge, MA, 2006)
2. R. Caruana, Multitask learning. Mach. Learn. **28**(1), 41–75 (1997)
3. M. Wainwright, *High-Dimensional Statistics: A Non-Asymptotic Viewpoint* (Cambridge University Press, Cambridge, 2019)
4. P. Bühlmann, S. van de Geer, *Statistics for High-Dimensional Data* (Springer, New York, 2011)
5. S. Shalev-Shwartz, S. Ben-David, *Understanding Machine Learning—From Theory to Algorithms* (Cambridge University Press, Cambridge, 2014)
6. V.N. Vapnik, *The Nature of Statistical Learning Theory* (Springer, Berlin, 1999)
7. S. Boyd, L. Vandenberghe, *Convex Optimization* (Cambridge University Press, Cambridge, UK, 2004)
8. D.P. Bertsekas, *Nonlinear Programming*, 2nd edn. (Athena Scientific, Belmont, MA, 1999)
9. T. Hastie, R. Tibshirani, J. Friedman, *The Elements of Statistical Learning* Springer Series in Statistics. (Springer, New York, 2001)
10. T. Hastie, R. Tibshirani, M. Wainwright, *Statistical Learning with Sparsity: The Lasso and Its Generalizations* (CRC Press, Boca Raton, FL, 2015)
11. A. Jung, A fixed-point of view on gradient methods for big data. Frontiers in Applied Mathematics and Statistics **3**, 18 (2017)
12. Y.C. Eldar, P. Kuppinger, H. Bölcskei, Block-sparse signals: Uncertainty relations and efficient recovery. IEEE Trans. Signal Processing **58**(6), 3042–3054 (2010). (June)
13. S. Pan, Q. Yang, A survey on transfer learning. IEEE Trans. Knowl. Data Eng. **22**(10), 1345–1359 (2010)
14. J. Howard, S. Ruder, Universal language model fine-tuning for text classification, in *Proceedings of the 56th Annual Meeting of the Association for Computational Linguistics (Volume 1: Long Papers)* (Association for Computational Linguistics, Stroudsburg, 2018), pp. 328–339

Chapter 8
Clustering

So far we focused on ML methods that use the ERM principle and lean a hypothesis by minimizing the discrepancy between its predictions and the true labels on a training set. These methods are referred to as supervised methods as they require labeled data points for which the true label values have been determined by some human (who serves as a "supervisor"). This and the following chapter discuss ML methods which do not require any labeled data point. These methods are often referred to as "unsupervised" since they do not require a supervisor to provide the label values for any data point.

The basic idea of clustering is that the data points arising in a ML application can be decomposed into few subsets which we refer to as **clusters**. Clustering methods learn a hypothesis for assigning each data point either to one cluster (see Sect. 8.1) or several clusters with different degrees of belonging (see Sect. 8.2). Two data points are assigned to the same cluster if they are similar to each other. Different clustering methods use different measures for the "similarity" between data points. For data points characterized by (numeric) Euclidean feature vectors, the similarity between data points can be naturally defined in terms of the Euclidean distance between feature vectors. Section 8.3 discusses clustering methods that use notions of similarity which do not require to characterize data points by Euclidean feature vectors (Fig. 8.1).

There is a strong conceptual link between clustering methods and the classification methods discussed in Chap. 3. Both type of methods learn a hypothesis that reads in the features of a data point an outputs a prediction for some quantity of interest. In classification methods, this quantity of interest is some generic label of a data point. For clustering methods, this quantity of interest is the index of the cluster to which a data point belongs to. A main difference between clustering and classification is that clustering methods do not require knowledge of the true label (cluster index) of any data point.

Classification methods learn a good hypothesis via minimizing their average loss incurred on a training set of labeled data points. In contrast, clustering methods do not have access to a single labeled data point. To find the correct labels (cluster

© The Author(s), under exclusive license to Springer Nature Singapore Pte Ltd. 2022   153
A. Jung, *Machine Learning*, Machine Learning: Foundations, Methodologies,
and Applications, https://doi.org/10.1007/978-981-16-8193-6_8

Fig. 8.1 Each circle represents an image which is characterized by its average redness x_r and average greenness x_g. The ith image is depicted by a circle located at the point $\mathbf{x}^{(i)} = \left(x_r^{(i)}, x_g^{(i)}\right)^T \in \mathbb{R}^2$. It seems that the images can be grouped into two clusters

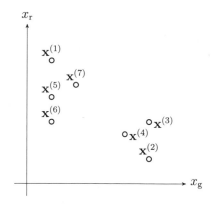

assignments) clustering methods rely solely on the intrinsic geometry of the data points. We will see that clustering methods use this intrinsic geometry to define an empirical risk incurred by a candidate hypothesis. Like classification methods, also clustering methods use an instance of the ERM principle (see Chap. 4) to find a good hypothesis (clustering).

This chapter discusses two main flavours of clustering methods:

- hard clustering (see Sect. 8.1)
- and soft clustering methods (see Sect. 8.2).

Hard clustering methods learn a hypothesis h that reads in the feature vector $\mathbf{x}$ of a data point and delivers a predicted cluster index $\hat{y} = h(\mathbf{x}) \in \{1, \ldots, k\}$. Thus, hard clustering assigns each data point to one single cluster. Section 8.1 will discuss one of the most widely-used hard clustering algorithms which is known as k-means.

In contrast to hard clustering methods, soft clustering methods assign each data point to several clusters with different degrees of belonging. These methods learn a hypothesis that delivers a vector $\hat{\mathbf{y}} = \left(\hat{y}_1, \ldots, \hat{y}_k\right)^T$ with entry $\hat{y}_c \in [0, 1]$ being the predicted degree of the data point belonging to the cluster with index c. Hard clustering is an extreme case of soft-clustering with requiring degrees of belonging taking values in $\{0, 1\}$ and allowing only one of them to be non-zero.

The main focus of this chapter is on methods that require data points being represented by numeric feature vectors (see Sects. 8.1 and 8.2). These methods define the similarity between data points using the Euclidean distance between their feature vectors. Some applications generate data points for which it is not obvious how to obtain numeric feature vectors such that their Euclidean distances reflect the similarity between data points. It is then desirable to use a more flexible notion of similarity which does not require to determine (useful) numeric feature vectors of data points. Maybe the most fundamental concept to represent similarities between data points is a similarity graph. The nodes of the similarity graph are the individual data points of a dataset. Similar data points are connected by edges (links) that might be assigned some weight that quantifies the amount of similarity. Section 8.3 discusses clustering methods that use a graph to represent similarities between data points.

8.1   Hard Clustering with k-Means

Consider a dataset $\mathcal{D}$ which consists of m data points that are indexed by $i = 1, \ldots, m$. We can access the data points only via their numeric feature vectors $\mathbf{x}^{(i)} \in \mathbb{R}^n$, for $i = 1, \ldots, m$. It will be convenient for the following discussion if we identify a data point with its feature vector. In particular, we refer by $\mathbf{x}^{(i)}$ to the ith data point. Hard clustering methods decompose (or cluster) the dataset into a given number k of different clusters $\mathcal{C}^{(1)}, \ldots, \mathcal{C}^{(k)}$. Hard clustering assigns each data point $\mathbf{x}^{(i)}$ to one and only one cluster $\mathcal{C}^{(c)}$ with the cluster index $c \in \{1, \ldots, k\}$.

Let us define for each data point its label $y^{(i)} \in \{1, \ldots, k\}$ as the index of the cluster to which the ith data point actually belongs to. The cth cluster consists of all data points with $y^{(i)} = c$,

$$\mathcal{C}^{(c)} := \{i \in \{1, \ldots, m\} : y^{(i)} = c\}. \tag{8.1}$$

We can interpret hard clustering methods as methods that compute predictions $\hat{y}^{(i)}$ for the cluster ("correct") assignments $y^{(i)}$. The predicted cluster assignments result in the predicted clusters

$$\widehat{\mathcal{C}}^{(c)} := \{i \in \{1, \ldots, m\} : \hat{y}^{(i)} = c\}, \text{ for } c = 1, \ldots, k. \tag{8.2}$$

We now discuss a widely used clustering method, known as k-means. This method does not require the knowledge of the label or (true) cluster assignment $y^{(i)}$ for any data point in $\mathcal{D}$. This method computes predicted cluster assignments $\hat{y}^{(i)}$ based solely from the intrinsic geometry of the feature vectors $\mathbf{x}^{(i)} \in \mathbb{R}^n$ for all $i = 1, \ldots, m$. Since it does not require any labeled data points, k-means is often referred to as being an unsupervised method. However, note that k-means requires the number k of clusters to be given as an input (or hyper-) parameter.

The k-means method represents the cth cluster $\widehat{\mathcal{C}}^{(c)}$ by a representative feature vector $\boldsymbol{\mu}^{(c)} \in \mathbb{R}^n$. It seems reasonable to assign data points in $\mathcal{D}$ to clusters $\widehat{\mathcal{C}}^{(c)}$ such that they are well concentrated around the cluster representatives $\boldsymbol{\mu}^{(c)}$. We make this informal requirement precise by defining the **clustering error**

$$\widehat{L}\big(\{\boldsymbol{\mu}^{(c)}\}_{c=1}^k, \{\hat{y}^{(i)}\}_{i=1}^m \mid \mathcal{D}\big) = (1/m) \sum_{i=1}^m \left\| \mathbf{x}^{(i)} - \boldsymbol{\mu}^{(\hat{y}^{(i)})} \right\|^2. \tag{8.3}$$

Note that the clustering error $\widehat{L}$ (8.3) depends on both, the cluster assignments $\hat{y}^{(i)}$, which define the cluster (8.2), and the cluster representatives $\boldsymbol{\mu}^{(c)}$, for $c = 1, \ldots, k$.

Finding the optimal cluster means $\{\boldsymbol{\mu}^{(c)}\}_{c=1}^k$ and cluster assignments $\{\hat{y}^{(i)}\}_{i=1}^m$ that minimize the clustering error (8.3) is computationally challenging. The difficulty stems from the fact that the clustering error is a non-convex function of the cluster means and assignments. While jointly optimizing the cluster means and assignments is hard, separately optimizing either the cluster means for given assignments or

Fig. 8.2 The flow of
k-means. Starting from an
initial guess or estimate for
the cluster means, the cluster
assignments and cluster
means are updated
(improved) in an alternating
fashion

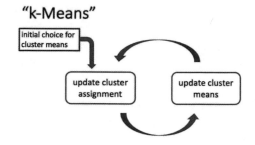

vice-versa is easy. In what follows, we present simple closed-form solutions for
these sub-problems. The k-means method simply combines these solutions in an
alternating fashion.

It can be shown that for given predictions (cluster assignments) $\hat{y}^{(i)}$, the clustering
error (8.3) is minimized by setting the cluster representatives equal to the **cluster
means** [1]

$$\boldsymbol{\mu}^{(c)} := \left(1/|\widehat{\mathcal{C}}^{(c)}|\right) \sum_{\hat{y}^{(i)}=c} \mathbf{x}^{(i)}. \tag{8.4}$$

To evaluate (8.4) we need to know the predicted cluster assignments $\hat{y}^{(i)}$. The crux
is that the optimal predictions $\hat{y}^{(i)}$, in the sense of minimizing clustering error (8.3),
depend themselves on the choice for the cluster representatives $\boldsymbol{\mu}^{(c)}$. In particular, for
given cluster representative $\boldsymbol{\mu}^{(c)}$ with $c = 1, \ldots, k$, the clustering error is minimized
by the cluster assignments

$$\hat{y}^{(i)} \in \underset{c \in \{1,\ldots,k\}}{\operatorname{argmin}} \left\| \mathbf{x}^{(i)} - \boldsymbol{\mu}^{(c)} \right\|. \tag{8.5}$$

Here, we denote by $\operatorname{argmin}_{c' \in \{1,\ldots,k\}} \|\mathbf{x}^{(i)} - \boldsymbol{\mu}^{(c')}\|$ the set of all cluster indices $c \in$
$\{1, \ldots, k\}$ such that $\|\mathbf{x}^{(i)} - \boldsymbol{\mu}^{(c)}\| = \min_{c' \in \{1,\ldots,k\}} \|\mathbf{x}^{(i)} - \boldsymbol{\mu}^{(c')}\|$.

Note that (8.5) assigns the ith data point to those cluster $\mathcal{C}^{(c)}$ whose cluster mean
$\boldsymbol{\mu}^{(c)}$ is nearest (in Euclidean distance) to $\mathbf{x}^{(i)}$. Thus, if we knew the optimal cluster
representatives, we could predict the cluster assignments using (8.5). However, we do
not know the optimal cluster representatives unless we have found good predictions
for the cluster assignments $\hat{y}^{(i)}$ (see (8.4)).

To recap: We have characterized the optimal choice (8.4) for the cluster representatives for given cluster assignments and the optimal choice (8.5) for the cluster
assignments for given cluster representatives. It seems natural, starting from some initial guess for the cluster representatives, to alternate between the cluster assignment
update (8.5) and the update (8.4) for the cluster means. This alternating optimization strategy is illustrated in Fig. 8.2 and summarized in Algorithm 12. Note that
Algorithm 12, which is maybe the most basic variant of k-means, simply alternates
between the two updates (8.4) and (8.5) until some stopping criterion is satisfied.

Algorithm 12 "k-means"

Input: dataset $\mathcal{D} = \{\mathbf{x}^{(i)}\}_{i=1}^m$; number k of clusters; initial cluster means $\boldsymbol{\mu}^{(c)}$ for $c = 1, \ldots, k$.

1: **repeat**
2:   for each data point $\mathbf{x}^{(i)}$, $i = 1, \ldots, m$, do

$$\hat{y}^{(i)} := \operatorname{argmin}_{c' \in \{1, \ldots, k\}} \|\mathbf{x}^{(i)} - \boldsymbol{\mu}^{(c')}\| \quad \text{(update cluster assignments)} \qquad (8.6)$$

3:   for each cluster $c = 1, \ldots, k$ do

$$\boldsymbol{\mu}^{(c)} := \frac{1}{|\{i : \hat{y}^{(i)} = c\}|} \sum_{i:\hat{y}^{(i)}=c} \mathbf{x}^{(i)} \quad \text{(update cluster means)} \qquad (8.7)$$

4: **until** stopping criterion is met
5: compute final clustering error $E^{(k)} := (1/m) \sum_{i=1}^m \left\|\mathbf{x}^{(i)} - \boldsymbol{\mu}^{(\hat{y}^{(i)})}\right\|^2$

Output: cluster means $\boldsymbol{\mu}^{(c)}$, for $c = 1, \ldots, k$, cluster assignments $\hat{y}^{(i)} \in \{1, \ldots, k\}$, for $i = 1, \ldots, m$, final clustering error $E^{(k)}$

Algorithm 12 requires the specification of the number k of clusters and initial choices for the cluster means $\boldsymbol{\mu}^{(c)}$, for $c = 1, \ldots, k$. Those quantities are hyperparameters that must be tuned to the specific geometry of the given dataset $\mathcal{D}$. This tuning can be based on probabilistic models for the dataset and its cluster structure (see Sect. 2.1.4 and [2, 3]). Alternatively, if Algorithm 12 is used as pre-processing within an overall supervised ML method (see Chap. 3), the validation error (see Sect. 6.3) of the overall method might guide the choice of the number k of clusters.

Choosing Number of Clusters. The choice for the number k of clusters typically depends on the role of the clustering method within an overall ML application. If the clustering method serves as a pre-processing for a supervised ML problem, we could try out different values of the number k and determine, for each choice k, the corresponding validation error. We then pick the value of k which results in the smallest validation error. If the clustering method is mainly used as a tool for data visualization, we might prefer a small number of clusters. The choice for the number k of clusters can also be guided by the so-called "elbow-method". Here, we run the k-means Algorithm 12 for several different choices of k. For each value of k, Algorithm 12 delivers a clustering with clustering error

$$E^{(k)} = \widehat{L}\big(\{\boldsymbol{\mu}^{(c)}\}_{c=1}^k, \{\hat{y}^{(i)}\}_{i=1}^m \mid \mathcal{D}\big).$$

We then plot the minimum empirical error $E^{(k)}$ as a function of the number k of clusters. Figure 8.3 depicts an example for such a plot which typically starts with a steep decrease for increasing k and then flattening out for larger values of k. Note that for $k \geq m$ we can achieve zero clustering error since each data point $\mathbf{x}^{(i)}$ can be assigned to a separate cluster $\mathcal{C}^{(c)}$ whose mean coincides with that data point, $\mathbf{x}^{(i)} = \boldsymbol{\mu}^{(c)}$.

Cluster-Means Initialization. We briefly mention some popular strategies for choosing the initial cluster means in Algorithm 12. One option is to initialize the

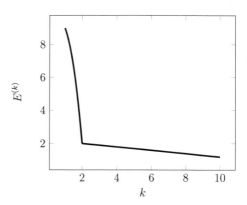

Fig. 8.3 The clustering error $E^{(k)}$ achieved by k-means for increasing number k of clusters

cluster means with realizations of i.i.d. random vectors whose probability distribution is matched to the dataset $\mathcal{D} = \{\mathbf{x}^{(i)}\}_{i=1}^{m}$ (see Sect. 3.12). For example, we could use a multivariate normal distribution $\mathcal{N}(\mathbf{x}; \widehat{\boldsymbol{\mu}}, \widehat{\boldsymbol{\Sigma}})$ with the sample mean $\widehat{\boldsymbol{\mu}} = (1/m)\sum_{i=1}^{m}\mathbf{x}^{(i)}$ and the sample covariance $\widehat{\boldsymbol{\Sigma}} = (1/m)\sum_{i=1}^{m}(\mathbf{x}^{(i)} - \widehat{\boldsymbol{\mu}})(\mathbf{x}^{(i)} - \widehat{\boldsymbol{\mu}})^{T}$. Alternatively, we could choose the initial cluster means $\boldsymbol{\mu}^{(c)}$ by selecting k different data points $\mathbf{x}^{(i)}$ from $\mathcal{D}$. This selection process might combine random choices with an optimization of the distances between cluster means [4]. Finally, the cluster means might also be chosen by evenly partitioning the principal component of the dataset (see Chap. 9).

Interpretation as ERM. For a practical implementation of Algorithm 12 we need to decide when to stop updating the cluster means and assignments (see (8.6) and (8.7). To this end it is useful to interpret Algorithm 12 as a method for iteratively minimizing the clustering error (8.3). As can be verified easily, the updates (8.6) and (8.7) always modify (update) the cluster means or assignments in such a way that the clustering error (8.3) is never increased. Thus, each new iteration of Algorithm 12 results in cluster means and assignments with a smaller (or the same) clustering error compared to the cluster means and assignments obtained after the previous iteration. Algorithm 12 implements a form of ERM (see Chap. 4) using the clustering error (8.3) as the empirical risk incurred by the predicted cluster assignments $\hat{y}^{(i)}$. Note that after completing a full iteration of Algorithm 12, the cluster means $\left\{\boldsymbol{\mu}^{(c)}\right\}_{c=1}^{k}$ are fully determined by the cluster assignments $\left\{\hat{y}^{(i)}\right\}_{i=1}^{m}$ via (8.7). It seems natural to terminate Algorithm 12 if the decrease in the clustering error achieved by the most recent iteration is below a prescribed (small) threshold.

Clustering and Classification. There is a strong conceptual link between Algorithm 12 and classification methods (see e.g. Sect. 3.13). Both methods essentially learn a hypothesis $h(\mathbf{x})$ that maps the feature vector $\mathbf{x}$ to a predicted label $\hat{y} = h(\mathbf{x})$ from a finite set. The practical meaning of the label values is different for Algorithm 12 and classification methods. For classification methods, the meaning of the label values is essentially defined by the training set (of labeled data points) used for ERM (4.3). On the other hand, clustering methods use the predicted label $\hat{y} = h(\mathbf{x})$ as a cluster index.

Another main difference between Algorithm 12 and most classification methods is the choice for the empirical risk used to evaluate the quality or usefulness of a given hypothesis $h(\cdot)$. Classification methods typically use an average loss over labeled data points in a training set as empirical risk. In contrast, Algorithm 12 uses the clustering error (8.3) as a form of empirical risk. Consider a hypothesis that resembles the cluster assignments $\hat{y}^{(i)}$ obtained after completing an iteration in Algorithm 12, $\hat{y}^{(i)} = h(\mathbf{x}^{(i)})$. Then we can rewrite the resulting clustering error achieved after this iteration as

$$\widehat{L}(h|\mathcal{D}) = (1/m) \sum_{i=1}^{m} \left\| \mathbf{x}^{(i)} - \frac{\sum_{i':h(\mathbf{x}^{(i)})=h(\mathbf{x}^{(i')})} \mathbf{x}^{(i')}}{\sum_{i':h(\mathbf{x}^{(i)})=h(\mathbf{x}^{(i')})}} \right\|^2. \tag{8.8}$$

Note that the ith summand in (8.8) depends on the entire dataset $\mathcal{D}$ and not only on the feature vector $\mathbf{x}^{(i)}$.

Some Practicalities. For a practical implementation of Algorithm 12 we need to fix three issues.

- Issue 1 ("tie-breaking"): We need to specify what to do if several different cluster indices $c \in \{1, \ldots, k\}$ achieve the minimum value in the cluster assignment update (8.6) during step 2.
- Issue 2 ("empty cluster"): The cluster assignment update (8.6) in step 3 of Algorithm 12 might result in a cluster c with no data points associated with it, $|\{i : \hat{y}^{(i)} = c\}| = 0$. For such a cluster c, the update (8.7) is not well-defined.
- Issue 3 ("stopping criterion"): We need to specify a criterion used in step 4 of Algorithm 12 to decide when to stop iterating.

Algorithm 13 is obtained from Algorithm 12 by fixing those three issues [5]. Step 3 of Algorithm 13 solves the first issue mentioned above ("tie breaking"), arising when there are several cluster clusters whose means have minimum distance to a data point $\mathbf{x}^{(i)}$, by assigning $\mathbf{x}^{(i)}$ to the cluster with smallest cluster index (see (8.9)). Step 4 of Algorithm 13 resolves the "empty cluster" issue by computing the variables $b^{(c)} \in \{0, 1\}$ for $c = 1, \ldots, k$. The variable $b^{(c)}$ indicates if the cluster with index c is active ($b^{(c)} = 1$) or the cluster c is inactive ($b^{(c)} = 0$). The cluster c is defined to be inactive if there are no data points assigned to it during the preceding cluster assignment step (8.9). The cluster activity indicators $b^{(c)}$ allows to restrict the cluster mean updates (8.10) only to the clusters c with at least one data point $\mathbf{x}^{(i)}$. To obtain a stopping criterion, step 7 Algorithm 13 monitors the clustering error $E^{(r)}$ incurred by the cluster means and assignments obtained after r iterations. Algorithm 13 continues updating cluster assignments (8.9) and cluster means (8.10) as long as the decrease is above a given threshold $\varepsilon \geq 0$.

For Algorithm 13 to be useful we must ensure that the stopping criterion is met within a finite number of iterations. In other words, we must ensure that the clustering error decrease can be made arbitrarily small within a sufficiently large (but finite) number of iterations. To this end, it is useful to represent Algorithm 13 as a fixed-point iteration

Algorithm 13 "k-Means II" (slight variation of "Fixed Point Algorithm" in [5])

Input: dataset $\mathcal{D} = \{\mathbf{x}^{(i)}\}_{i=1}^{m}$; number k of clusters; tolerance $\varepsilon \geq 0$; initial cluster means $\{\boldsymbol{\mu}^{(c)}\}_{c=1}^{k}$

1: **Initialize.** set iteration counter $r := 0$; $E^{(0)} := 0$

2: **repeat**

3:      for all data points $i = 1, \ldots, m$,

$$\hat{y}^{(i)} := \min\{\text{argmin}_{c' \in \{1,\ldots,k\}} \|\mathbf{x}^{(i)} - \boldsymbol{\mu}^{(c')}\|\} \quad \text{(update cluster assignments)} \qquad (8.9)$$

4:      for all clusters $c = 1, \ldots, k$, update the activity indicator

$$b^{(c)} := \begin{cases} 1 & \text{if } |\{i : \hat{y}^{(i)} = c\}| > 0 \\ 0 & \text{else.} \end{cases}$$

5:      for all $c = 1, \ldots, k$ with $b^{(c)} = 1$,

$$\boldsymbol{\mu}^{(c)} := \frac{1}{|\{i : \hat{y}^{(i)} = c\}|} \sum_{\{i:\hat{y}^{(i)}=c\}} \mathbf{x}^{(i)} \quad \text{(update cluster means)} \qquad (8.10)$$

6:      $r := r + 1$ (increment iteration counter)

7:      $E^{(r)} := \widehat{L}\big(\{\boldsymbol{\mu}^{(c)}\}_{c=1}^{k}, \{\hat{y}^{(i)}\}_{i=1}^{m} \mid \mathcal{D}\big)$ (evaluate clustering error (8.3))

8: **until** $r > 1$ and $E^{(r-1)} - E^{(r)} \leq \varepsilon$ (check for sufficient decrease in clustering error)

9: $E^{(k)} := (1/m) \sum_{i=1}^{m} \left\|\mathbf{x}^{(i)} - \boldsymbol{\mu}^{(\hat{y}^{(i)})}\right\|^{2}$ (compute final clustering error)

Output: cluster assignments $\hat{y}^{(i)} \in \{1, \ldots, k\}$, cluster means $\boldsymbol{\mu}^{(c)}$, clustering error $E^{(k)}$.

$$\{\hat{y}^{(i)}\}_{i=1}^{m} \mapsto \mathcal{P}\{\hat{y}^{(i)}\}_{m=1}^{m}. \qquad (8.11)$$

The operator $\mathcal{P}$, which depends on the dataset $\mathcal{D}$, reads in a list of cluster assignments and delivers an improved list of cluster assignments aiming at reducing the associated clustering error (8.3). Each iteration of Algorithm 13 updates the cluster assignments $\hat{y}^{(i)}$ by applying the operator $\mathcal{P}$. Representing Algorithm 13 as a fixed-point iteration (8.11) allows for an elegant proof of the convergence of Algorithm 13 within a finite number of iterations (even for $\varepsilon = 0$) [5, Thm. 2].

Figure 8.4 depicts the evolution of the cluster assignments and cluster means during the iterations Algorithm 13. Each subplot corresponds to one iteration of Algorithm 13 and depicts the cluster means before that iteration and the clustering assignments (via the marker symbols) after the corresponding iteration. In particular, the upper left subplot depicts the cluster means before the first iteration (which are the initial cluster means) and the cluster assignments obtained after the first iteration of Algorithm 13.

Consider running Algorithm 13 with tolerance $\varepsilon = 0$ (see step 8) such that the iterations are continued until there is no decrease in the clustering error $E^{(r)}$ (see step 7 of Algorithm 13). As discussed above, Algorithm 13 will terminate after a

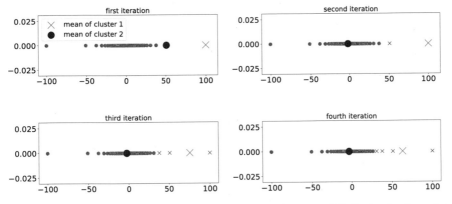

Fig. 8.4 The evolution of cluster means (8.7) and cluster assignments (8.6) (depicted as large dot and large cross) during the first four iterations of k-means Algorithm 13

finite number of iterations. Moreover, for $\varepsilon = 0$, the delivered cluster assignments $\{\hat{y}^{(i)}\}_{i=1}^{m}$ are fully determined by the delivered clustered means $\{\boldsymbol{\mu}^{(c)}\}_{c=1}^{k}$,

$$\hat{y}^{(i)} = \min\{ \underset{c' \in \{1,\ldots,k\}}{\arg\min} \|\mathbf{x}^{(i)} - \boldsymbol{\mu}^{(c')}\| \}. \tag{8.12}$$

Indeed, if (8.12) does not hold one can show the final iteration r would still decrease the clustering error and the stopping criterion in step 8 would not be met.

If cluster assignments and cluster means satisfy the condition (8.12), we can rewrite the clustering error (8.3) as a function of the cluster means solely,

$$\widehat{L}(\{\boldsymbol{\mu}^{(c)}\}_{c=1}^{k} | \mathcal{D}) := (1/m) \sum_{i=1}^{m} \min_{c' \in \{1,\ldots,k\}} \|\mathbf{x}^{(i)} - \boldsymbol{\mu}^{(c')}\|^2. \tag{8.13}$$

Even for cluster assignments and cluster means that do not satisfy (8.12), we can still use (8.13) to lower bound the clustering error (8.3),

$$\widehat{L}(\{\boldsymbol{\mu}^{(c)}\}_{c=1}^{k} | \mathcal{D}) \leq \widehat{L}(\{\boldsymbol{\mu}^{(c)}\}_{c=1}^{k}, \{\hat{y}^{(i)}\}_{i=1}^{m} | \mathcal{D}).$$

Algorithm 13 iteratively improves the cluster means in order to minimize (8.13). Ideally, we would like Algorithm 13 to deliver cluster means that achieve the global minimum of (8.13) (see Fig. 8.5). However, for some combination of dataset $\mathcal{D}$ and initial cluster means, Algorithm 13 delivers cluster means that form only a local optimum of $\widehat{L}(\{\boldsymbol{\mu}^{(c)}\}_{c=1}^{k} | \mathcal{D})$ which is strictly worse (larger) than its global optimum (see Fig. 8.5).

The tendency of Algorithm 13 to get trapped around a local minimum of (8.13) depends on the initial choice for cluster means. Therefore, it is often useful to repeat Algorithm 13 several times, with each repetition using a different initial choice for

Fig. 8.5 The clustering error (8.13) is a non-convex function of the cluster means $\{\boldsymbol{\mu}^{(c)}\}_{c=1}^{k}$. Algorithm 13 iteratively updates cluster means to minimize the clustering error but might get trapped around one of its local minimum

the cluster means. We then pick the cluster assignments $\{\hat{y}^{(i)}\}_{i=1}^{m}$ obtained for the repetition that resulted in the smallest clustering error $E^{(k)}$ (see step 9).

8.2 Soft Clustering with Gaussian Mixture Models

Consider a dataset $\mathcal{D} = \{\mathbf{x}^{(1)}, \ldots, \mathbf{x}^{(m)}\}$ that we wish to group into a given number of k different clusters. The hard clustering methods of Sect. 8.1 deliver (predicted) cluster assignments $\hat{y}^{(i)}$ as the index of the cluster to which data point $\mathbf{x}^{(i)}$ is assigned to. These cluster assignments $\hat{y}$ provide rather coarse-grained information. Two data points $\mathbf{x}^{(i)}$, $\mathbf{x}^{(j)}$ might be assigned to the same cluster c although their distances to the cluster mean $\boldsymbol{\mu}^{(c)}$ might be very different. Intuitively, these two data points have a different degree of belonging to the cluster c.

For some clustering applications it is desirable to quantify the degree by which a data point belongs to a cluster. Soft clustering methods use a continues range, such as the closed interval $[0, 1]$, of possible values for the degree of belonging. In contrast, hard clustering methods use only two possible degrees of belonging, either full belonging or no belonging to a cluster. While hard clustering methods assign a given data point to precisely one cluster, soft clustering methods typically assign a data point to several different clusters with non-zero degree of belonging.

This chapter discusses soft clustering methods that compute, for each data point $\mathbf{x}^{(i)}$ in the dataset $\mathcal{D}$, a vector $\widehat{\mathbf{y}}^{(i)} = \left(\hat{y}_{1}^{(i)}, \ldots, \hat{y}_{k}^{(i)}\right)^{T}$. We can interpret the entry $\hat{y}_{c}^{(i)} \in [0, 1]$ as the degree by which the data point $\mathbf{x}^{(i)}$ belongs to the cluster $\mathcal{C}^{(c)}$. For $\hat{y}_{c}^{(i)} \approx 1$, we are quite confident in the data point $\mathbf{x}^{(i)}$ belonging to cluster $\mathcal{C}^{(c)}$. In contrast, for $\hat{y}_{c}^{(i)} \approx 0$, we are quite confident in the data point $\mathbf{x}^{(i)}$ being outside the cluster $\mathcal{C}^{(c)}$.

A widely used soft-clustering method uses a probabilistic model for the data points $\mathcal{D} = \{\mathbf{x}^{(i)}\}_{i=1}^{m}$. Within this model, each cluster $\mathcal{C}^{(c)}$, for $c = 1, \ldots, k$, is represented by a multivariate normal distributions [6]

$$\mathcal{N}(\mathbf{x}; \boldsymbol{\mu}^{(c)}, \boldsymbol{\Sigma}^{(c)}) = \frac{1}{\sqrt{\det\{2\pi\boldsymbol{\Sigma}\}}} \exp\left(-(1/2)(\mathbf{x}-\boldsymbol{\mu}^{(c)})^T(\boldsymbol{\Sigma}^{(c)})^{-1}(\mathbf{x}-\boldsymbol{\mu}^{(c)})\right), \text{ for } c = 1, \dots, k.$$

(8.14)

The probability distribution (8.14) is parameterized by a cluster-specific mean vector $\boldsymbol{\mu}^{(c)}$ and an (invertible) cluster-specific covariance matrix $\boldsymbol{\Sigma}^{(c)}$.[1] We interpret a specific data point $\mathbf{x}^{(i)}$ as a realization drawn from the probability distribution (8.14) of a specific cluster $c^{(i)}$,

$$\mathbf{x}^{(i)} \sim \mathcal{N}(\mathbf{x}; \boldsymbol{\mu}^{(c)}, \boldsymbol{\Sigma}^{(c)}) \text{ with cluster index } c = c^{(i)}.$$

(8.15)

We can think of $c^{(i)}$ as the true index of the cluster to which the data point $\mathbf{x}^{(i)}$ belongs to. The variable $c^{(i)}$ selects the cluster distributions (8.14) from which the feature vector $\mathbf{x}^{(i)}$ has been generated (drawn). We will therefore refer to the variable $c^{(i)}$ as the (true) cluster assignment for the ith data point. Similar to the feature vectors $\mathbf{x}^{(i)}$ we also interpret the cluster assignments $c^{(i)}$, for $i = 1, \dots, m$ as realizations of i.i.d. RVs.

In contrast to the feature vectors $\mathbf{x}^{(i)}$, we do not observe (know) the true cluster indices $c^{(i)}$. After all, the goal of soft clustering is to estimate the cluster indices $c^{(i)}$. We obtain a soft-clustering method by estimating the cluster indices $c^{(i)}$ based solely on the data points in $\mathcal{D}$. To compute these estimates we assume that the (true) cluster indices $c^{(i)}$ are realizations of i.i.d. RVs with the common probability distribution (or probability mass function)

$$p_c := p(c^{(i)} = c) \text{ for } c = 1, \dots, k.$$

(8.16)

The (prior) probabilities p_c, for $c = 1, \dots, k$, are either assumed known or estimated from data [6, 7]. The choice for the probabilities p_c could reflect some prior knowledge about different sizes of the clusters. For example, if cluster $\mathcal{C}^{(1)}$ is known to be larger than cluster $\mathcal{C}^{(2)}$, we might choose the prior probabilities such that $p_1 > p_2$.

The probabilistic model given by (8.15), (8.16) is referred to as a a GMM. This name is quite natural as the common marginal distribution for the feature vectors $\mathbf{x}^{(i)}$, for $i = 1, \dots, m$, is a (additive) mixture of multivariate normal (Gaussian) distributions,

$$p(\mathbf{x}^{(i)}) = \sum_{c=1}^{k} \underbrace{p(c^{(i)} = c)}_{p_c} \underbrace{p(\mathbf{x}^{(i)}|c^{(i)} = c)}_{\mathcal{N}(\mathbf{x}^{(i)};\boldsymbol{\mu}^{(c)},\boldsymbol{\Sigma}^{(c)})}.$$

(8.17)

As already mentioned, the cluster assignments $c^{(i)}$ are hidden (unobserved) RVs. We thus have to infer or estimate these variables from the observed data points $\mathbf{x}^{(i)}$ which realizations or i.i.d. RVs with the common distribution (8.17).

---

[1] Note that the expression (8.14) is only valid for an invertible (non-singular) covariance matrix $\boldsymbol{\Sigma}$.

The GMM (see (8.15) and (8.16)) lends naturally to a rigorous definition for the degree $y_c^{(i)}$ by which data point $\mathbf{x}^{(i)}$ belongs to cluster c.[2] Let us define the label value $y_c^{(i)}$ as the "a-posteriori" probability of the cluster assignment $c^{(i)}$ being equal to $c \in \{1, \ldots, k\}$:

$$y_c^{(i)} := p(c^{(i)} = c | \mathcal{D}). \tag{8.18}$$

By their very definition (8.18), the degrees of belonging $y_c^{(i)}$ always sum to one,

$$\sum_{c=1}^{k} y_c^{(i)} = 1 \text{ for each } i = 1, \ldots, m. \tag{8.19}$$

We emphasize that we use the conditional cluster probability (8.18), conditioned on the dataset $\mathcal{D}$, for defining the degree of belonging $y_c^{(i)}$. This is reasonable since the degree of belonging $y_c^{(i)}$ depends on the overall (cluster) geometry of the data set $\mathcal{D}$.

The definition (8.18) for the label values (degrees of belonging) $y_c^{(i)}$ involves the GMM parameters $\{\boldsymbol{\mu}^{(c)}, \boldsymbol{\mu}^{(c)}, p_c\}_{c=1}^{k}$. Since we do not know these parameters beforehand we cannot evaluate the conditional probability in (8.18). A principled approach to solve this problem is to evaluate (8.18) with the true GMM parameters replaced by some estimates $\{\widehat{\boldsymbol{\mu}}^{(c)}, \widehat{\boldsymbol{\Sigma}}^{(c)}, \hat{p}_c\}_{c=1}^{k}$. Plugging in the GMM parameter estimates into (8.18) provides us with predictions $\hat{y}_c^{(i)}$ for the degrees of belonging. However, to compute the GMM parameter estimates we would have already needed the degrees of belonging $y_c^{(i)}$. This situation is similar to hard clustering where ultime goals is to jointly optimize cluster means and assignments (see Sect. 8.1).

Similar to the spirit of Algorithm 4 for hard clustering, we solve the above dilemma of soft clustering by an alternating optimization scheme. This scheme alternates between updating (optimizing) the predicted degrees of belonging $\hat{y}_c^{(i)}$, for $i = 1, \ldots, m$ and $c = 1, \ldots, k$, given the current GMM parameter estimates $\{\widehat{\boldsymbol{\mu}}^{(c)}, \widehat{\boldsymbol{\Sigma}}^{(c)}, \hat{p}_c\}_{c=1}^{k}$ and then updating (optimizing) these GMM parameter estimates based on the updated predictions $\hat{y}_c^{(i)}$. We summarize the resulting soft clustering method in Algorithm 4. Each iteration of Algorithm 4 consists of an update (8.22) for the degrees of belonging followed by an update (step 3) for the GMM parameters (Fig. 8.6).

To analyze Algorithm 14 it is helpful to interpret (the features of) data points $\mathbf{x}^{(i)}$ as realizations of i.i.d. RVs distributed according to a GMM (8.15)–(8.16). We can then understand Algorithm 14 as a method for estimating the GMM parameters based on observing realizations drawn from the GMM (8.15)–(8.16). A principled approach to estimating the parameters of a probability distribution is the maximum likelihood method (see Sect. 3.12 and [7, 8]). The idea is to estimate the GMM parameters by maximizing the probability (density)

---

[2] Remember that the degree of belongings $y_c^{(i)}$ are considered as (unknown) label values for datapoints. The choice or definition for the labels of data points is a design choice. In particular, we can define the labels of data points using a hypothetical probabilistic model such as the GMM.

Fig. 8.6 The GMM (8.15), (8.16) yields a probability distribution (8.17) which is a weighted sum of multivariate normal distributions $\mathcal{N}(\boldsymbol{\mu}^{(c)}, \boldsymbol{\Sigma}^{(c)})$. The weight of the cth component is the cluster probability $p(c^{(i)} = c)$

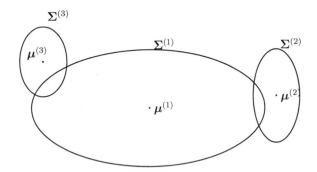

$$p\left(\mathcal{D}; \{\boldsymbol{\mu}^{(c)}, \boldsymbol{\Sigma}^{(c)}, p_c\}_{c=1}^{k}\right) \qquad (8.20)$$

of actually observing the data point in the dataset $\mathcal{D}$.

It can be shown that Algorithm 14 is an instance of a generic approximate maximum likelihood technique referred to as expectation maximization expectation maximization (EM) (see [9, Chap. 8.5] for more details). In particular, each iteration of Algorithm 14 updates the GMM parameter estimates such that the corresponding probability density (8.20) does not decrease [10]. If we denote the GMM parameter estimate obtained after r iterations of Algorithm 14 by $\boldsymbol{\theta}^{(r)}$ [9, Sect. 8.5.2],

$$p\left(\mathcal{D}; \boldsymbol{\theta}^{(r+1)}\right) \ge p\left(\mathcal{D}; \boldsymbol{\theta}^{(r)}\right) \qquad (8.21)$$

Algorithm 14 "A Soft-Clustering Algorithm" [1]

Input: dataset $\mathcal{D} = \{\mathbf{x}^{(i)}\}_{i=1}^{m}$; number k of clusters, initial GMM parameter estimates $\{\widehat{\boldsymbol{\mu}}^{(c)}, \widehat{\boldsymbol{\Sigma}}^{(c)}, \hat{p}_c\}_{c=1}^{k}$

1: **repeat**

2:      for each $i = 1, \ldots, m$ and $c = 1, \ldots, k$, update degrees of belonging

$$\hat{y}_c^{(i)} := \frac{\hat{p}_c \mathcal{N}(\mathbf{x}^{(i)}; \widehat{\boldsymbol{\mu}}^{(c)}, \widehat{\boldsymbol{\Sigma}}^{(c)})}{\sum_{c'=1}^{k} \hat{p}_{c'} \mathcal{N}(\mathbf{x}^{(i)}; \widehat{\boldsymbol{\mu}}^{(c')}, \widehat{\boldsymbol{\Sigma}}^{(c')})} \qquad (8.22)$$

3:      for each $c \in \{1, \ldots, k\}$, update GMM parameter estimates:

- $\hat{p}_c := m_c/m$ with effective cluster size $m_c := \sum_{i=1}^{m} \hat{y}_c^{(i)}$ (cluster probability)

- $\widehat{\boldsymbol{\mu}}^{(c)} := (1/m_c) \sum_{i=1}^{m} \hat{y}_c^{(i)} \mathbf{x}^{(i)}$ (cluster mean)

- $\widehat{\boldsymbol{\Sigma}}^{(c)} := (1/m_c) \sum_{i=1}^{m} \hat{y}_c^{(i)} \left(\mathbf{x}^{(i)} - \widehat{\boldsymbol{\mu}}^{(c)}\right)\left(\mathbf{x}^{(i)} - \widehat{\boldsymbol{\mu}}^{(c)}\right)^T$ (cluster covariance matrix)

4: **until** stopping criterion met

Output: predicted degrees of belonging $\widehat{\mathbf{y}}^{(i)} = (\hat{y}_1^{(i)}, \ldots, \hat{y}_k^{(i)})^T$ for $i = 1, \ldots, m$.

As for Algorithm 12, we can also interpret Algorithm 4 as an instance of the ERM principle discussed in Chap. 4. Indeed, maximizing the probability density (8.20) is equivalent to minimizing the empirical risk

$$\widehat{L}(\boldsymbol{\theta} \mid \mathcal{D}) := -\log p(\mathcal{D}; \boldsymbol{\theta}) \text{ with GMM parameters } \boldsymbol{\theta} := \{\boldsymbol{\mu}^{(c)}, \boldsymbol{\Sigma}^{(c)}, p_c\}_{c=1}^k \tag{8.23}$$

The empirical risk (8.23) is the negative logarithm of the probability (density) (8.20) of observing the dataset $\mathcal{D}$ as i.i.d. realizations of the GMM (8.17). The monotone increase in the probability density (8.21) achieved by the iterations of Algorithm 4 translate into a monotone decrease of the empirical risk,

$$\widehat{L}(\boldsymbol{\theta}^{(r)} \mid \mathcal{D}) \leq \widehat{L}(\boldsymbol{\theta}^{(r-1)} \mid \mathcal{D}) \text{ with iteration counter } r. \tag{8.24}$$

The monotone decrease (8.24) in the empirical risk (8.23) achieved by the iterations of Algorithm 14 naturally lends to a stopping criterion. Let $E^{(r)}$ denote the empirical risk (8.23) achieved by the GMM parameter estimates $\boldsymbol{\theta}^{(r)}$ obtained after r iterations in Algorithm 14. We stop iterating as soon as the decrease $E^{(r)} - E^{(r+1)}$ achieved by the $r + 1$th iteration of Algorithm 4 falls below a given (positive) threshold $\varepsilon > 0$.

Similar to Algorithm 12, also Algorithm 14 might get trapped in local minima of the underlying empirical risk. The GMM parameters delivered by Algorithm 14 might only be a local minimum of (8.23) but not the global minimum (see Fig. 8.5 for the analogous situation in hard clustering). As for hard clustering Algorithm 12, we typically repeat Algorithm 14 several times. During each repetition of Algorithm 14, we use a different (randomly chosen) initialization for the GMM parameter estimates $\boldsymbol{\theta} = \{\widehat{\boldsymbol{\mu}}^{(c)}, \widehat{\boldsymbol{\Sigma}}^{(c)}, \hat{p}_c\}_{c=1}^k$. Each repetition of Algorithm 14 results in a potentially different set of GMM parameter estimates and degrees of belongings $\hat{y}_c^{(i)}$. We then use the results for that repetition that achieves the smallest empirical risk (8.23).

Let us point out an interesting link between soft clustering methods based on GMM (see Algorithm 14) and hard clustering with k-means (see Algorithm 12). Consider the GMM (8.15) with prescribed cluster covariance matrices

$$\boldsymbol{\Sigma}^{(c)} = \sigma^2 \mathbf{I} \text{ for all } c \in \{1, \ldots, k\}, \tag{8.25}$$

with some given variance $\sigma^2 > 0$. We assume the cluster covariance matrices in the GMM to be given by (8.25) and therefore can replace the covariance matrix updates in Algorithm 14 with the assignment $\widehat{\boldsymbol{\Sigma}}^{(c)} := \sigma^2 \mathbf{I}$. It can be verified easily that for sufficiently small variance σ^2 in (8.25), the update (8.22) tends to enforce $\hat{y}_c^{(i)} \in \{0, 1\}$. In other words, each data point $\mathbf{x}^{(i)}$ becomes then effectively associated with exactly one single cluster c whose cluster mean $\widehat{\boldsymbol{\mu}}^{(c)}$ is nearest to $\mathbf{x}^{(i)}$. For $\sigma^2 \to 0$, the soft-clustering update (8.22) in Algorithm 14 reduces to the (hard) cluster assignment update (8.6) in k-means Algorithm 12. We can interpret Algorithm 12 as an extreme case of Algorithm 14 that is obtained by fixing the covariance matrices in the GMM to $\sigma^2 \mathbf{I}$ with a sufficiently small σ^2.

Combining GMM with linear regression. Let us now sketch how Algorithm 14 could be combined with linear regression methods. The idea is to first compute the degree of belongings to the clusters for each data point. We then learn separate linear predictors for each cluster using the degree of belongings as weights for the individual loss terms in the training error. To predict the label of a new data point, we first compute the predictions obtained for each cluster-specific linear hypothesis and then average them using the degree of the new data point belonging to each cluster.

8.3 Connectivity-Based Clustering

The clustering methods discussed in Sects. 8.1 and 8.2 can only be applied to data points which are characterized by numeric feature vectors. These methods define the similarity between data points using the Euclidean distance between the feature vectors of these data points. As illustrated in Fig. 8.7, these methods can only produce "Euclidean shaped" clusters that are contained either within hyper-spheres (Algorithm 12) or hyper-ellipsoids (Algorithm 14).

Some applications generate data points for which the construction of useful numeric features is difficult. Even if we can easily obtain numeric features for data points, the Euclidean distances between the resulting feature vectors might not reflect the actual similarities between data points. As a case in point, consider data points representing text documents. We could use the histogram of a respecified list of words as numeric features for a text document. In general, a small Euclidean distance between histograms of text documents does not imply that the text documents have similar meanings. Moreover, groups or clusters of similar text documents might have highly complicated shapes in the space of feature vectors that cannot be grouped within hyper-ellipsoids. For datasets with such "non-Euclidean" cluster shapes, k-means or GMM are not suitable as clustering methods. We should then replace the Euclidean distance between feature vectors with another concept to determine or measure the similarity between data points.

Connectivity-based clustering methods do not require any numeric features of data points. These methods cluster data points based on explicitly specifying for

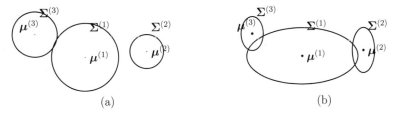

(a)   (b)

Fig. 8.7 **a** Cartoon of typical cluster shapes delivered by k-means Algorithm 13. **b** Cartoon of typical cluster shapes delivered by soft clustering Algorithm 14

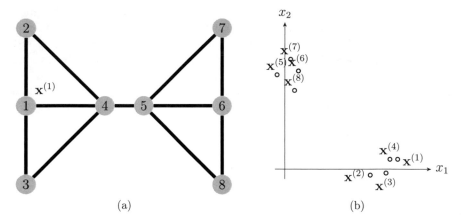

(a)   (b)

Fig. 8.8 Connectivity-based clustering can be obtained by constructing features $\mathbf{x}^{(i)}$ that are (approximately) identical for well-connected data points. **a** A similarity graph for a dataset $\mathcal{D}$ consists of nodes representing individual data points and edges that connect similar data points. **b** Feature vectors of well-connected data points have small Euclidean distance

any two different data points if they are similar and to what extend. A convenient mathematical tool to represent similarities between the data points of a dataset $\mathcal{D}$ is a weighted undirected graph $\mathcal{G} = (\mathcal{V}, \mathcal{E})$. We refer to this graph as the similarity graph of the dataset $\mathcal{D}$ (see Fig. 8.8). The nodes $\mathcal{V}$ in this similarity graph $\mathcal{G}$ represent data points in $\mathcal{D}$ and the undirected edges connect nodes that represent similar data points. The extend of the similarity is represented by the weights $W_{i,j}$ for each edge $\{i, j\} \in \mathcal{E}$.

Given a similarity graph $\mathcal{G}$ of a dataset, connectivity-based clustering methods determine clusters as subsets of nodes that are well connected within the cluster but weakly connected between different clusters. Different concepts for quantifying the connectivity between nodes in a graph yield different clustering methods. Spectral clustering methods use eigenvectors of a graph Laplacian matrix to measure the connectivity between nodes [11, 12]. Flow-based clustering methods measure the connectivity between two nodes via the amount of flow that can be routed between them [13]. Note that we might use these connectivity measures to construct meaningful numerical feature vectors for the nodes in the empirical graph. These feature vectors can then be fed into the hard-clustering Algorithm 13 or the soft clustering Algorithm 14 (see Fig. 8.8).

The density-based clustering algorithm DBSCAN considers two data points i, i' as connected if one of them (say i) is a core node and the other node (i') can be reached via a sequence (path) of connected core nodes

$$i^{(1)}, \ldots, i^{(r)}, \text{ with } \{i, i^{(1)}\}, \{i^{(1)}, i^{(2)}\}, \ldots, \{i^{(r)}, i'\} \in \mathcal{E}.$$

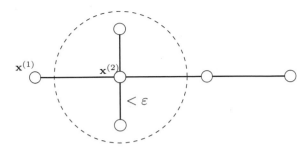

Fig. 8.9 DBSCAN assigns two data points to the same cluster if they are reachable. Two data points $\mathbf{x}^{(i)}, \mathbf{x}^{(i')}$ are reachable if there is a path of data points from $\mathbf{x}^{(i')}$ to $\mathbf{x}^{(i)}$. This path consists of a sequence of data points that are within a distance of ε. Moreover, each data point on this path must be a core point which has at least a given number of neighbouring data points within the distance ε

DBSCAN considers a node to be a core node if it has a sufficiently large number of neighbours [14]. The minimum number of neighbours required for a node to be considered a core node is a hyperparameter of DBSCAN. When DBSCAN is applied to data points with numeric feature vectors, it defined two data points as similar if the Euclidean distance between their feature vectors does not exceed a given threshold ε (see Fig. 8.9).

In contrast to k-means and GMM, DBSCAN does not require the number of clusters to be specified. The number of clusters is determined automatically by DBSCAN and depends on its hyperparameters. DBSCAN also performs an implicit outlier detection. The outliers delivered by DBSCAN are those data points which do not belong to the same cluster as another data point.

8.4 Clustering as Preprocessing

In applications it might be beneficial to combine clustering methods with supervised methods such as linear regression. As a point in case consider a dataset that consists of data points obtained from two different data generation processes. Let us denote the data points generated by one process by $\mathcal{D}^{(1)}$ and the other one by $\mathcal{D}^{(2)}$. Each datapoint is characterized by features and a label. While there would be an accurate linear hypothesis for predicting the label of data points in $\mathcal{D}^{(1)}$ and another linear hypothesis for $\mathcal{D}^{(2)}$ these two arc very different.

We could try to use clustering methods to assign any given data point to the corresponding data generation process. If we are lucky, the resulting clusters resemble (approximately) the two data generation processes $\mathcal{D}^{(1)}$ and $\mathcal{D}^{(2)}$. Once we have successfully clustered the data points, we can learn a separate (tailored) hypothesis for ach cluster. More generally, we can use the predicted cluster assignments obtained from the methods of Sects. 8.1–8.3 as additional features for each data point.

Let us illustrate the above ideas by combining Algorithm 12 with linear regression. We first group data points into a given number k of clusters and then learn separate linear predictors $h^{(c)}(\mathbf{x}) = \left(\mathbf{w}^{(c)}\right)^T \mathbf{x}$ for each cluster $c = 1, \ldots, k$. To predict the label of a new data point with features $\mathbf{x}$, we first assign to the cluster c' with the nearest cluster mean. We then use the linear predictor $h^{(c')}$ assigned to cluster c' to compute the predicted label $\hat{y} = h^{(c')}(\mathbf{x})$.

8.5 Exercises

Exercise 8.1 Monoticity of k-means **Updates** Show that the cluster means and assignments updates (8.7) and (8.6) never increase the clustering error (8.3).

Exercise 8.2 How to choose k **in** k-means? Discuss strategies for choosing the number k of clusters which is used as a hyper-parameter for k-means.

Exercise 8.3 Local Minima. Consider applying the hard clustering Algorithm 13 to the dataset $(-10, 1), (10, 1), (-10, -1), (10, -1)$ with initial cluster means $(0, 1), (0, -1)$ and tolerance $\varepsilon = 0$. For this initialization, will Algorithm 13 get trapped in a local minimum of the clustering error (8.13)?

Exercise 8.4 Image Compression with k-means Apply k-means to image compression. Consider image pixels as data points whose features are RGB intensities. We obtain a simple image compression format by, instead of storing RGB pixel values, storing the cluster means (which are RGB triplets) and the cluster index for each pixel.

Exercise 8.5 Compression with k-means Consider $m = 10000$ data points with feature vectors $\mathbf{x}^{(1)}, \ldots, \mathbf{x}^{(m)}$ length two. We apply k-means to cluster the data set into two clusters. How many bits do we need to store the clustering? For simplicity, we assume that any real number can be stored perfectly as a floating point numbers (32 bit).

References

1. C.M. Bishop, *Pattern Recognition and Machine Learning* (Springer, Berlin, 2006)
2. B. Kulis, M.I. Jordan, Revisiting k-means: new algorithms via bayesian nonparametrics, in *Proceedings of the 29th International Conference on Machine Learning, ICML 2012, Edinburgh, Scotland, UK, June 26 - July 1, 2012*. icml.cc/Omnipress (2012)
3. S. Wade, Z. Ghahramani, Bayesian cluster analysis: point estimation and credible balls (with discussion). Bayesian Anal. **13**(2), 559–626 (2018)
4. D. Arthur, S. Vassilvitskii, k-means++: the advantages of careful seeding, in *Proceedings of the Eighteenth Annual ACM-SIAM Symposium on Discrete Algorithms* (Society for Industrial and Applied Mathematics, Philadelphia, 2007)

5. R. Gray, J. Kieffer, Y. Linde, Locally optimal block quantizer design. Inf. Control **45**, 178–198 (1980)
6. D. Bertsekas, J. Tsitsiklis, *Introduction to Probability*. Athena Scientific, 2 edn. (2008)
7. E.L. Lehmann, G. Casella, *Theory of Point Estimation*, 2nd edn. (Springer, New York, 1998)
8. S.M. Kay, *Fundamentals of Statistical Signal Processing: Estimation Theory* (Prentice Hall, Englewood Cliffs, 1993)
9. T. Hastie, R. Tibshirani, J. Friedman, *The Elements of Statistical Learning* Springer Series in Statistics. (Springer, New York, 2001)
10. L. Xu, M. Jordan, On convergence properties of the EM algorithm for Gaussian mixtures. Neural Comput. **8**(1), 129–151 (1996)
11. U. von Luxburg, A tutorial on spectral clustering. Stat. Comput. **17**(4), 395–416 (2007). (Dec.)
12. A.Y. Ng, M.I. Jordan, Y. Weiss, On spectral clustering: analysis and an algorithm, in *Advances in Neural Information Processing Systems* (2001)
13. A. Jung, Y. SarcheshmehPour, Local graph clustering with network lasso. IEEE Signal Process. Lett. **28**, 106–110 (2021)
14. M. Ester, H.-P. Kriegel, J. Sander, X. Xu, A density-based algorithm for discovering clusters a density-based algorithm for discovering clusters in large spatial databases with noise, in *Proceedings of the Second International Conference on Knowledge Discovery and Data Mining*. Portland, Oregon (1996), pp. 226–231

Chapter 9
Feature Learning

"Solving Problems By Changing the Viewpoint."

Chapter 2 discussed features as those properties of a data point that can be measured or computed easily. Sometimes the choice of features follows naturally from the available hard and software. For example, we might use the numeric measurement $z \in \mathbb{R}$ delivered by a sensing device as a feature. However, we could augment this single feature with new features such as the powers z^2 and z^3 or adding a constant $z + 5$. Each of these computations produces a new feature. Which of these additional features are most useful?

Feature learning methods automate the choice of finding good features. These methods learn a hypothesis map that reads in some representation of a data point and transforms it to a set of features. Feature learning methods differ in the precise format of the original data representation as well as the format of the delivered features. The focus of this chapter in on feature learning methods that require data points being represented by d numeric raw features and deliver a set of n new numeric features. We will denote the set of raw and new features by $\mathbf{z} = (z_1, \ldots, z_d)^T \in \mathbb{R}^d$ and $\mathbf{x} = (x_1, \ldots, x_n)^T \in \mathbb{R}^n$, respectively.

Many ML application domains generate data points for which can access a huge number of raw features. Consider data points being snapshots generated by a smartphone. It seems natural to use the pixel colour intensities as the raw features of the snapshot. Since modern smartphone have Megapixel cameras, the pixel intensities would provide us with millions of raw features. It might seem a good idea to use as many (raw) features of a data point as possible since more features should offer more information about a data point and its label y. There are, however, two pitfalls in using an unnecessarily large number of features. The first one is a computational pitfall and the second one is a statistical pitfall.

Computationally, using very large feature vectors $\mathbf{x} \in \mathbb{R}^n$ (with n being billions), might result in excessive resource requirements (bandwidth, storage, time) of the

A. Jung, *Machine Learning*, Machine Learning: Foundations, Methodologies, and Applications, https://doi.org/10.1007/978-981-16-8193-6_9

resulting ML method. Statistically, using a large number of features makes the resulting ML methods more prone to overfitting. For example, linear regression will typically overfit when using feature vectors $\mathbf{x} \in \mathbb{R}^n$ whose length n exceeds the number m of labeled data points used for training (see Chap. 7).

Both from a computational and a statistical perspective, it is beneficial to use only the maximum necessary amount of features. The challenge is to select those features which carry most of the relevant information required for the prediction of the label y. Finding the most relevant features out of a huge number of (raw) features is the goal of dimensionality reduction methods. Dimensionality reduction methods form an important sub-class of feature learning methods. Formally, dimensionality reduction methods learn a hypothesis $h(\mathbf{z})$ that map a long raw feature vector $\mathbf{z} \in \mathbb{R}^d$ to a short new feature vector $\mathbf{x} \in \mathbb{R}^n$ with $d \gg n$.

Beside avoiding overfitting and coping with limited computational resources, dimensionality reduction can also be useful for data visualization. Indeed, if the resulting feature vector has length $n = 2$, we depict data points in the two-dimensional plane in form of a scatterplot.

We will discuss the basic idea underlying dimensionality reduction methods in Sect. 9.1. Section 9.2 presents one particular example of a dimensionality reduction method that computes relevant features by a linear transformation of the raw feature vector. Section 9.4 discusses a method for dimensionality reduction that exploits the availability of labelled data points. Section 9.6 shows how randomness can be used to obtain computationally cheap dimensionality reduction.

Most of this chapter discusses dimensionality reduction methods that determine a small number of relevant features from a large set of raw features. However, sometimes it might be useful to go the opposite direction. There are applications where it might be beneficial to construct a large (even infinite) number of new features from a small set of raw features. Section 9.7 will showcase how computing additional features can help to improve the prediction accuracy of ML methods.

9.1   Basic Principle of Dimensionality Reduction

The efficiency of ML methods depends crucially on the choice of features that are used to characterize data points. Ideally we would like to have a small number of highly relevant features to characterize data points. If we use too many features we risk to waste computations on exploring irrelevant features. If we use too few features we might not have enough information to predict the label of a data point. For a given number n of features, dimensionality reduction methods aim at learning an (in a certain sense) optimal map from the data point to a feature vector of length n.

Figure 9.1 illustrates the basic idea of dimensionality reduction methods. Their goal is to learn (or find) a "compression" map $h(\cdot) : \mathbb{R}^d \to \mathbb{R}^n$ that transforms a (long) raw feature vector $\mathbf{z} \in \mathbb{R}^d$ to a (short) feature vector $\mathbf{x} = (x_1, \ldots, x_n)^T := h(\mathbf{z})$ (typically $n \ll d$).

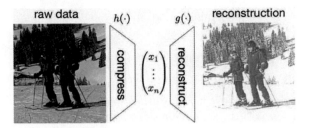

Fig. 9.1 Dimensionality reduction methods aim at finding a map h which maximally compresses the raw data while still allowing to accurately reconstruct the original data point from a small number of features $x_1, \ldots, x_n$

The new feature vector $\mathbf{x} = h(\mathbf{z})$ serves as a compressed representation (or code) for the original raw feature vector $\mathbf{z}$. We can reconstruct the raw feature vector using a reconstruction map $r(\cdot) : \mathbb{R}^n \to \mathbb{R}^d$. The reconstructed raw features $\widehat{\mathbf{z}} := r(\mathbf{x}) = r(h(\mathbf{z}))$ will typically by different from the original raw feature vector $\mathbf{z}$. Thus, we will obtain a non-zero reconstruction error

$$\underbrace{\widehat{\mathbf{z}}}_{=r(h(\mathbf{z})))} - \mathbf{z}. \tag{9.1}$$

Dimensionality reduction methods learn a compression map $h(\cdot)$ such that the reconstruction error (9.1) is minimized. In particular, for a dataset $\mathcal{D} = \{\mathbf{z}^{(1)}, \ldots, \mathbf{z}^{(m)}\}$, we measure the quality of a pair of compression map h and reconstruction map r by the average reconstruction error

$$\widehat{L}(h, r | \mathcal{D}) := (1/m) \sum_{i=1}^{m} L(\mathbf{z}^{(i)}, r(h(\mathbf{z}^{(i)}))). \tag{9.2}$$

Here, $L(\mathbf{z}, r(h(\mathbf{z}^{(i)})))$ denotes a loss function that is used to measure the reconstruction error $\underbrace{r(h(\mathbf{z}^{(i)}))}_{\widehat{\mathbf{z}}} - \mathbf{z}$. Different choices for the loss function in (9.2) result in different dimensionality reduction methods. One widely-used choice for the loss is the squared Euclidean norm

$$L(\mathbf{z}, g(h(\mathbf{z}))) := \|\mathbf{z} - g(h(\mathbf{z}))\|_2^2. \tag{9.3}$$

Practical dimensionality reduction methods have only finite computational resources. Any practical method must therefore restrict the set of possible compression and reconstruction maps to small subsets $\mathcal{H}$ and $\mathcal{H}^*$, respectively. These subsets are the hypothesis spaces for the compression map $h \in \mathcal{H}$ and the reconstruction map $r \in \mathcal{H}^*$. Feature learning methods differ in their choice for these hypothesis spaces.

Dimensionality reduction methods learn a compression map by solving

$$\hat{h} = \underset{h \in \mathcal{H}}{\arg\min} \underset{r \in \mathcal{H}^*}{\min} \widehat{L}(h, r | \mathcal{D})$$

$$\overset{(9.2)}{=} \underset{h \in \mathcal{H}}{\arg\min} \underset{r \in \mathcal{H}^*}{\min} (1/m) \sum_{i=1}^{m} L(\mathbf{z}^{(i)}, r(h(\mathbf{z}^{(i)}))). \qquad (9.4)$$

We can interpret (9.4) as a (typically non-linear) approximation problem. The optimal compression map $\hat{h}$ is such that the reconstruction $r(\hat{h}(\mathbf{z}))$, with a suitably chosen reconstruction map r, approximates the original raw feature vector $\mathbf{z}$ as good as possible. Note that we use a single compression map $h(\cdot)$ and a single reconstruction map $r(\cdot)$ for all data points in the dataset $\mathcal{D}$.

We obtain variety of dimensionality methods by using different choices for the hypothesis spaces $\mathcal{H}$, $\mathcal{H}^*$ and loss function in (9.4). Section 9.2 discusses a method that solves (9.4) for $\mathcal{H}$, $\mathcal{H}^*$ constituted by linear maps and the loss (9.3). Deep autoencoders are another family of dimensionality reduction methods that solve (9.4) with $\mathcal{H}$, $\mathcal{H}^*$ constituted by non-linear maps that are represented by deep neural networks [1, Chap. 14].

9.2 Principal Component Analysis

We now consider the special case of dimensionality reduction where the compression and reconstruction map are required to be linear maps. Consider a data point which is characterized by a (typically very long) raw feature vector $\mathbf{z} \in \mathbb{R}^d$ of length d. The length d of the raw feature vector might be easily of the order of millions. To obtain a small set of relevant features $\mathbf{x} = (x_1, \ldots, x_n)^T \in \mathbb{R}^n$, we apply a linear transformation to the raw feature vector,

$$\mathbf{x} = \mathbf{W}\mathbf{z}. \qquad (9.5)$$

Here, the "compression" matrix $\mathbf{W} \in \mathbb{R}^{n \times d}$ maps (in a linear fashion) the (long) raw feature vector $\mathbf{z} \in \mathbb{R}^d$ to the (shorter) feature vector $\mathbf{x} \in \mathbb{R}^n$.

It is reasonable to choose the compression matrix $\mathbf{W} \in \mathbb{R}^{n \times D}$ in (9.5) such that the resulting features $\mathbf{x} \in \mathbb{R}^n$ allow to approximate the original data point $\mathbf{z} \in \mathbb{R}^d$ as accurate as possible. We can approximate (or recover) the data point $\mathbf{z} \in \mathbb{R}^d$ back from the features $\mathbf{x}$ by applying a reconstruction operator $\mathbf{R} \in \mathbb{R}^{d \times n}$, which is chosen such that

$$\mathbf{z} \approx \mathbf{R}\mathbf{x} \overset{(9.5)}{=} \mathbf{R}\mathbf{W}\mathbf{z}. \qquad (9.6)$$

The approximation error $\widehat{L}(\mathbf{W}, \mathbf{R} | \mathcal{D})$ resulting when (9.6) is applied to each data point in a dataset $\mathcal{D} = \{\mathbf{z}^{(i)}\}_{i=1}^{m}$ is then

$$\widehat{L}(\mathbf{W}, \mathbf{R} | \mathcal{D}) = (1/m) \sum_{i=1}^{m} \|\mathbf{z}^{(i)} - \mathbf{R}\mathbf{W}\mathbf{z}^{(i)}\|^2. \qquad (9.7)$$

One can verify that the approximation error $\widehat{L}(\mathbf{W}, \mathbf{R} \mid \mathcal{D})$ can only by minimal if the compression matrix $\mathbf{W}$ is of the form

$$\mathbf{W} = \mathbf{W}_{\text{PCA}} := \left(\mathbf{u}^{(1)}, \ldots, \mathbf{u}^{(n)}\right)^T \in \mathbb{R}^{n \times d}, \tag{9.8}$$

with n orthonormal vectors $\mathbf{u}^{(j)}$, for $j = 1, \ldots, n$. The vectors $\mathbf{u}^{(j)}$ are the eigenvectors corresponding to the n largest eigenvalues of the sample covariance matrix

$$\mathbf{Q} := (1/m)\mathbf{Z}^T \mathbf{Z} \in \mathbb{R}^{d \times d}. \tag{9.9}$$

Here we used the data matrix $\mathbf{Z} = \left(\mathbf{z}^{(1)}, \ldots, \mathbf{z}^{(m)}\right)^T \in \mathbb{R}^{m \times d}$.[1] It can be verified easily, using the definition (9.9), that the matrix $\mathbf{Q}$ is psd. As a psd matrix, $\mathbf{Q}$ has an eigenvalue decomposition (EVD) of the form [2]

$$\mathbf{Q} = \left(\mathbf{u}^{(1)}, \ldots, \mathbf{u}^{(d)}\right) \begin{pmatrix} \lambda^{(1)} & \ldots & 0 \\ 0 & \ddots & 0 \\ 0 & \ldots & \lambda^{(d)} \end{pmatrix} \left(\mathbf{u}^{(1)}, \ldots, \mathbf{u}^{(d)}\right)^T$$

with real-valued eigenvalues $\lambda^{(1)} \geq \lambda^{(2)} \geq \ldots \geq \lambda^{(d)} \geq 0$ and orthonormal eigenvectors $\{\mathbf{u}_r\}_{r=1}^d$.

The feature vectors $\mathbf{x}^{(i)}$ are obtained by applying the compression matrix $\mathbf{W}_{\text{PCA}}$ (9.8) to the raw feature vectors $\mathbf{z}^{(i)}$. We refer to the entries of the vector $\mathbf{x}^{(i)}$, obtained via the eigenvectors of $\mathbf{Q}$ (see (9.2)), as the **principal components (PC)** of the raw feature vectors $\mathbf{z}^{(i)}$. Algorithm 15 summarizes the overall procedure of determining the compression matrix (9.8) and computing the vectors $\mathbf{x}^{(i)}$ whose entries are the PC of the raw feature vectors. This procedure is known as **principal component analysis (PCA)**. Note that the length $n(\leq d)$ of the mew feature vector $\mathbf{x}$, which is also the number of PCs used, is an input (or hyper) parameter of Algorithm 15. The number n can be chosen between the two extreme cases $n = 0$ (maximum compression) and $n = d$ (no compression). We finally note that the choice for the orthonormal eigenvectors in (9.8) might not be unique. Depending on the sample covariance matrix $\mathbf{Q}$, there might different sets of orthonormal vectors that correspond to the same eigenvalue of $\mathbf{Q}$. Thus, for a given length n of the new feature vectors, there might be several different matrices $\mathbf{W}$ that achieve the same (optimal) reconstruction error $\widehat{L}^{(\text{PCA})}$.

From a computational perspective, Algorithm 15 essentially amounts to performing an EVD of the sample covariance matrix $\mathbf{Q}$ (see (9.9)). Indeed, the EVD of $\mathbf{Q}$ provides not only the optimal compression matrix $\mathbf{W}_{\text{PCA}}$ but also the measure $\widehat{L}^{(\text{PCA})}$ for the information loss incurred by replacing the original data points $\mathbf{z}^{(i)} \in \mathbb{R}^d$ with the smaller feature vector $\mathbf{x}^{(i)} \in \mathbb{R}^n$. In particular, this information loss is measured by the approximation error (obtained for the optimal reconstruction matrix $\mathbf{R}_{\text{opt}} = \mathbf{W}_{\text{PCA}}^T$)

---

[1] Some authors define the data matrix as $\mathbf{Z} = \left(\widetilde{\mathbf{z}}^{(1)}, \ldots, \widetilde{\mathbf{z}}^{(m)}\right)^T \in \mathbb{R}^{m \times D}$ using "centered" raw feature vectors $\widetilde{\mathbf{z}}^{(i)} - \widehat{\mathbf{m}}$ obtained by subtracting the average $\widehat{\mathbf{m}} = (1/m) \sum_{i=1}^m \mathbf{z}^{(i)}$.

Algorithm 15 Principal Component Analysis (PCA)

Input: dataset $\mathcal{D} = \{\mathbf{z}^{(i)} \in \mathbb{R}^d\}_{i=1}^m$; number n of PCs.

1: compute the EVD (9.2) to obtain orthonormal eigenvectors $(\mathbf{u}^{(1)}, \ldots, \mathbf{u}^{(d)})$ corresponding to (decreasingly ordered) eigenvalues $\lambda^{(1)} \geq \lambda^{(2)} \geq \ldots \geq \lambda^{(d)} \geq 0$

2: construct compression matrix $\mathbf{W}_{\text{PCA}} := \left(\mathbf{u}^{(1)}, \ldots, \mathbf{u}^{(n)}\right)^T \in \mathbb{R}^{n \times d}$

3: compute feature vector $\mathbf{x}^{(i)} = \mathbf{W}_{\text{PCA}}\mathbf{z}^{(i)}$ whose entries are PC of $\mathbf{z}^{(i)}$

4: compute approximation error $\widehat{L}^{(\text{PCA})} = \sum_{r=n+1}^d \lambda^{(r)}$ (see (9.10)).

Output: $\mathbf{x}^{(i)}$, for $i = 1, \ldots, m$, and the approximation error $\widehat{L}^{(\text{PCA})}$.

Fig. 9.2 Reconstruction error $\widehat{L}^{(\text{PCA})}$ (see (9.10)) of PCA for varying number n of PCs

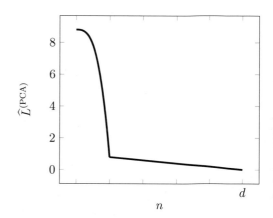

$$\widehat{L}^{(\text{PCA})} := \widehat{L}\big(\mathbf{W}_{\text{PCA}}, \underbrace{\mathbf{R}_{\text{opt}}}_{=\mathbf{W}_{\text{PCA}}^T} \mid \mathcal{D}\big) = \sum_{r=n+1}^d \lambda^{(r)}. \tag{9.10}$$

As depicted in Fig. 9.2, the approximation error $\widehat{L}^{(\text{PCA})}$ decreases with increasing number n of PCs used for the new features (9.5). For the extreme case $n = 0$, where we completely ignore the raw feature vectors $\mathbf{z}^{(i)}$, the optimal reconstruction error is $\widehat{L}^{(\text{PCA})} = (1/m) \sum_{i=1}^m \|\mathbf{z}^{(i)}\|^2$. The other extreme case $n = d$ allows to use the raw features directly as the new features $\mathbf{x}^{(i)} = \mathbf{z}^{(i)}$, which amounts to no compression at all, and trivially results in a zero reconstruction error $\widehat{L}^{(\text{PCA})} = 0$.

9.2.1   Combining PCA with Linear Regression

One important use case of PCA is as a pre-processing step within an overall ML problem such as linear regression (see Sect. 3.1). As discussed in Chap. 7, linear regression methods are prone to overfitting whenever the data points are characterized by feature vectors whose length D exceeds the number m of labeled data points used for training. One simple but powerful strategy to avoid overfitting is to preprocess the original feature vectors (they are considered as the raw data points $\mathbf{z}^{(i)} \in \mathbb{R}^d$) by applying PCA in order to obtain smaller feature vectors $\mathbf{x}^{(i)} \in \mathbb{R}^n$ with $n < m$.

9.2.2 How to Choose Number of PC?

There are several aspects which can guide the choice for the number n of PCs to be used as features.

- for data visualization: use either $n = 2$ or $n = 3$
- computational budget: choose n sufficiently small such that the computational complexity of the overall ML method does not exceed the available computational resources.
- statistical budget: consider using PCA as a pre-processing step within a linear regression problem (see Sect. 3.1). Thus, we use the output $\mathbf{x}^{(i)}$ of PCA as the feature vectors in linear regression. In order to avoid overfitting, we should choose $n < m$ (see Chap. 7).
- elbow method: choose n large enough such that approximation error $\widehat{L}^{(\mathrm{PCA})}$ is reasonably small (see Fig. 9.2).

9.2.3 Data Visualisation

If we use PCA with $n = 2$, we obtain feature vectors $\mathbf{x}^{(i)} = \mathbf{W}_{\mathrm{PCA}}\mathbf{z}^{(i)}$ (see (9.5)) which can be depicted as points in a scatterplot (see Sect. 2.1.3). As an example, consider data points $\mathbf{z}^{(i)}$ obtained from historic recordings of Bitcoin statistics. Each data point $\mathbf{z}^{(i)} \in \mathbb{R}^d$ is a vector of length $d = 6$. It is difficult to visualise points in an Euclidean space $\mathbb{R}^d$ of dimension $d > 2$. Therefore, we apply PCA with $n = 2$ which results in feature vectors $\mathbf{x}^{(i)} \in \mathbb{R}^2$. These new feature vectors (of length 2) can be depicted conveniently as a scatterplot (see Fig. 9.3).

9.2.4 Extensions of PCA

There have been proposed several extensions of the basic PCA method:

- **Kernel PCA** [3, Chap. 14.5.4]: The PCA method is most effective if the raw feature vectors of data points are nearby a low-dimensional linear subspace of $\mathbb{R}^d$. Kernel PCA extends PCA to handle data points that are located near a low-dimensional manifold which might be highly non-linear. This is achieved by applying PCA to transformed feature vectors instead of the original feature vectors. Kernel PCA first applies a (typically non-linear) feature map to the original feature vectors

Fig. 9.3 A scatterplot of
data points with feature
vectors $\mathbf{x}^{(i)} = \left(x_1^{(i)}, x_2^{(i)}\right)^T$
whose entries are the first
two PCs of the Bitcoin
statistics $\mathbf{z}^{(i)}$ of the ith day

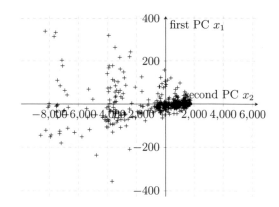

$\mathbf{x}^{(i)}$ resulting in new feature vectors $\mathbf{z}^{(i)}$ (see Sect. 3.9). We then apply PCA to the
transformed feature vectors $\mathbf{z}^{(i)}$, for $i = 1, \ldots, m$.

- **Robust PCA** [4]: In its basic form, PCA is sensitive to outliers which are a small
  number of data points with fundamentally different statistical properties than the
  bulk of data points. This sensitivity might be attributed to the properties of the
  squared Euclidean norm (9.3) which is used in PCA to measure the reconstruction
  error (9.1). We have seen in Chap. 3 that linear regression (see Sect. 3.1 and 3.3)
  can be made robust against outliers by replacing the squared error loss with another
  loss function. In a similar spirit, robust PCA replaces the squared Euclidean norm
  with another norm that is less sensitive to having very large reconstruction errors
  (9.1) for a small number of data points (which are outliers).

- **Sparse PCA** [3, Chap. 14.5.5]: The basic PCA method transforms the raw feature
  vector $\mathbf{z}^{(i)}$ of a data point to a new (shorter) feature vector $\mathbf{x}^{(i)}$. In general each entry
  $x_j^{(i)}$ of the new feature vectors will depend on every raw feature. More precisely, the
  new feature $x_j^{(i)}$ depends on all raw features $z_{j'}^{(i)}$ for which the corresponding entry
  $W_{j,j'}$ of the matrix $\mathbf{W} = \mathbf{W}_{\text{PCA}}$ (9.8) is non-zero. For most datasets, all entries of
  the matrix $\mathbf{W}_{\text{PCA}}$ will typically be non-zero.

  In some applications of linear dimensionality reduction we would like to construct
  new features that depend only on a small subset of raw features. Equivalently we
  would like to learn a linear compression map $\mathbf{W}$ (9.5) such that each row of $\mathbf{W}$
  contains only few non-zero entries. To this end, sparse PCA enforces the rows of
  the compression matrix $\mathbf{W}$ to contain only a small number of non-zero entries.
  This enforcement can be implement either using additional constraints on $\mathbf{W}$ or
  by adding a penalty term to the reconstruction error (9.7).

- **Probabilistic PCA** [5, 6]: We have motivated PCA as a method for learning an
  optimal linear compression map (matrix) (9.5) such that the compressed feature
  vectors allows to linearly reconstruct the original raw feature vector with minimum
  reconstruction error (9.7). Another interpretation of PCA is that of a method that

learns a subspace of $\mathbb{R}^d$ that best fits the set of raw feature vectors $\mathbf{z}^{(i)}$, for $i = 1, \ldots, m$. This optimal subspace is precisely the subspace spanned by the rows of $\mathbf{W}_{\text{PCA}}$ (9.8).

Probabilistic principal component analysis (PPCA) interprets the raw feature vectors $\mathbf{z}^{(i)}$ as realizations of i.i.d. RVs. These realizations are modelled as

$$\mathbf{z}^{(i)} = \mathbf{W}^T \mathbf{x}^{(i)} + \varepsilon^{(i)}, \text{ for } i = 1, \ldots, m. \tag{9.11}$$

Here, $\mathbf{W} \in \mathbb{R}^{n \times d}$ is some unknown matrix with orthonormal rows. The rows of $\mathbf{W}$ span the subspace around which the raw features are concentrated. The vectors $\mathbf{x}^{(i)}$ in (9.11) are realizations of i.i.d. RVs whose common probability distribution is $\mathcal{N}(\mathbf{0}, \mathbf{I})$. The vectors $\varepsilon^{(i)}$ are realizations of i.i.d. RVs whose common probability distribution is $\mathcal{N}(\mathbf{0}, \sigma^2 \mathbf{I})$ with some fixed but unknown variance σ^2. Note that $\mathbf{W}$ and σ^2 parametrize the joint probability distribution of the feature vectors $\mathbf{z}^{(i)}$ via (9.11). Probabilistic principal component analysis (PPCA) amounts to maximum likelihood estimation (see Sect. 3.12) of the parameters $\mathbf{W}$ and σ^2. This maximum likelihood estimation problem can be solved using computationally efficient estimation techniques such as EM [6, Appendix B]. The implementation of PPCA via EM also offers a principled approach to handle missing data. Roughly speaking, the EM method allows to use the probabilistic model (9.11) to estimate missing raw features [6, Sect. 4.1].

9.3 Feature Learning for Non-numeric Data

We have motivated dimensionality reduction methods as transformations of (very long) raw feature vectors to a new (shorter) feature vector $\mathbf{x}$ such that it allows to reconstruct $\mathbf{z}$ with minimum reconstruction error (9.1). To make this requirement precise we need to define a measure for the size of the reconstruction error and specify the class of possible reconstruction maps. PCA uses the squared Euclidean norm (9.7) to measure the reconstruction error and only allows for linear reconstruction maps (9.6).

Alternatively, we can view dimensionality reduction as the generation of new feature vectors $\mathbf{x}^{(i)}$ that maintain the intrinsic geometry of the data points with their raw feature vectors $\mathbf{z}^{(i)}$. Different dimensionality reduction methods using different concepts for characterizing the "intrinsic geometry" of data points. PCA defines the intrinsic geometry of data points using the squared Euclidean distances between feature vectors. Indeed, PCA produces feature vectors $\mathbf{x}^{(i)}$ such that for data points whose raw feature vectors have small squared Euclidean distance, also the new feature vectors $\mathbf{x}^{(i)}$ will have small squared Euclidean distance.

Some application domains generate data points for which the Euclidean distances between raw feature vectors does not reflect the intrinsic geometry of data points. As a point in case, consider data points representing scientific articles which can be characterized by the relative frequencies of words from some given set of relevant words (dictionary). A small Euclidean distance between the resulting raw feature vectors typically does not imply that the corresponding text documents are similar. Instead, the similarity between two articles might depend on the number of authors that are contained in author lists of both papers. We can represent the similarities between all articles using a similarity graph whose nodes represent data points which are connected by an edge (link) if they are similar (see Fig. 8.8).

Consider a dataset $\mathcal{D} = (\mathbf{z}^{(1)}, \ldots, \mathbf{z}^{(m)})$ whose intrinsic geometry is characterized by an unweighted similarity graph $\mathcal{G} = (\mathcal{V} := \{1, \ldots, m\} \, \mathcal{E})$. The node $i \in \mathcal{V}$ represents the ith data point, with raw feature vector $\mathbf{z}^{(i)}$. Two nodes are connected by an undirected edge if the corresponding data points are similar. We would like to find short feature vectors $\mathbf{x}^{(i)}$, for $i = 1, \ldots, m$, such that two data points i, i', whose feature vectors $\mathbf{x}^{(i)}, \mathbf{x}^{(i')}$ have small Euclidean distance, are well-connected to each other. To make this requirement precise we need to define a measure for how well two nodes of an undirected graph are connected. We refer the reader to literature on network theory for an overview and details of various connectivity measures [7].

Let us discuss a simple but powerful technique to map the nodes $i \in \mathcal{V}$ of an undirected graph $\mathcal{G}$ to (short) feature vectors $\mathbf{x}^{(i)} \in \mathbb{R}^n$. This map is such that the Euclidean distances between the feature vectors of two nodes reflect their connectivity within $\mathcal{G}$. This technique uses the Laplacian matrix $\mathbf{L} in \mathbb{R}^{(i)}$ which is defined for an undirected graph $\mathcal{G}$ (with node set $\mathcal{V} = \{1, \ldots, m\}$) element-wise

$$L_{i,j} := \begin{cases} -1 & \text{, if } \{i, j\} \in \mathcal{E} \\ d^{(i)} & \text{, if } i = j \\ 0 & \text{otherwise.} \end{cases} \tag{9.12}$$

Here, $d^{(i)} := \big|\{j : \{i, j\} \in \mathcal{E}\}\big|$ denotes the degree, or the number of neighbours, of node $i \in \mathcal{V}$. It can be shown that the Laplacian matrix $\mathbf{L}$ is psd [8, Proposition 1]. Therefore we can find an orthonormal set of eigenvectors

$$\mathbf{u}^{(1)}, \ldots, \mathbf{u}^{(m)} \in \mathbb{R}^m \tag{9.13}$$

with corresponding (ordered in a non-decreasing fashion) eigenvalues $\lambda_1 \leq \ldots \leq \lambda_m$ of $\mathbf{L}$.

It turns out that, for a prescribed number n of numeric features, the entries $u_i^{(1)}, \ldots, u_i^{(n)}$ of the first n eigenvectors (9.13) result in feature vectors whose Euclidean distances reflect the connectivities of data points in the similarity graph $\mathcal{G}$. For a more precise statement of this informal claim we refer to the excellent tutorial [8]. Thus, we obtain a feature learning method for (non-numeric) data points via using the eigenvectors of the graph Laplacian associated with the similarity graph of the data points. Algorithm 16 summarizes this feature learning method which requires the similarity graph of the dataset and the desired number of new features as input. Note that Algorithm 16 does not make any use of the Euclidean distances between raw feature vectors and uses solely the similarity graph $\mathcal{G}$ for determining the intrinsic geometry of $\mathcal{D}$.

Algorithm 16 Feature Learning for Non-Numeric Data

Input: dataset $\mathcal{D} = \{\mathbf{z}^{(i)} \in \mathbb{R}^d\}_{i=1}^m$; similarity graph $\mathcal{G}$; number n of features to be constructed for each data point.
1: construct the Laplacian matrix $\mathbf{L}$ of the similarity graph (see ((9.12)))
2: compute EVD of $\mathbf{L}$ to obtain n orthonormal eigenvectors (9.13) corresponding to the smallest eigenvalues of $\mathbf{L}$
3: for each data point i, construct feature vector

$$\mathbf{x}^{(i)} := \left(u_i^{(1)}, \ldots, u_i^{(n)}\right)^T \in \mathbb{R}^n \qquad (9.14)$$

Output: $\mathbf{x}^{(i)}$, for $i = 1, \ldots, m$

9.4 Feature Learning for Labeled Data

We have discussed PCA as a linear dimensionality reduction method. PCA learns a compression matrix that maps raw features $\mathbf{z}^{(i)}$ of data points to new (much shorter) feature vectors $\mathbf{x}^{(i)}$. The feature vectors $\mathbf{x}^{(i)}$ determined by PCA depend solely on the raw feature vectors $\mathbf{z}^{(i)}$ of the data points in a given dataset $\mathcal{D}$. In particular, PCA determines the compression matrix such that the new features allow for a linear reconstruction (9.6) with minimum reconstruction error (9.7).

For some application domains we might not only have access to raw feature vectors but also to the label values $y^{(i)}$ of the data points in $\mathcal{D}$. Indeed, dimensionality reduction methods might be used as pre-processing step within a regression or classification problem that involves a labeled training set. However, in its basic form, PCA (see Algorithm 15) does not allow to exploit the information provided by available labels $y^{(i)}$ of data points $\mathbf{z}^{(i)}$. For some datasets, PCA might deliver feature vectors that are not very relevant for the overall task of predicting the label of a data point.

Let us now discuss a modification of PCA that exploits the information provided by available labels of the data points. The idea is to learn a linear construction map (matrix) $\mathbf{W}$ such that the new feature vectors $\mathbf{x}^{(i)} = \mathbf{W}\mathbf{z}^{(i)}$ allow to predict the label $y^{(i)}$ as good as possible. We restrict the prediction to be linear,

$$\hat{y}^{(i)} := \mathbf{r}^T \mathbf{x}^{(i)} = \mathbf{r}^T \mathbf{W}\mathbf{z}^{(i)}, \qquad (9.15)$$

with some weight vector $\mathbf{r} \in \mathbb{R}^n$.

While PCA is motivated by minimizing the reconstruction error (9.1), we now aim at minimizing the prediction error $\hat{y}^{(i)} - y^{(i)}$. In particular, we assess the usefulness of a given pair of construction map $\mathbf{W}$ and predictor $\mathbf{r}$ (see (9.15)), using the empirical risk

$$\widehat{L}(\mathbf{W}, \mathbf{r} \mid \mathcal{D}) := (1/m) \sum_{i=1}^{m} \left(y^{(i)} - \hat{y}^{(i)}\right)^2$$

$$\overset{(9.15)}{=} (1/m) \sum_{i=1}^{m} \left(y^{(i)} - \mathbf{r}^T \mathbf{W} \mathbf{z}^{(i)}\right)^2. \tag{9.16}$$

to guide the learning of a compressing matrix $\mathbf{W}$ and corresponding linear predictor weights $\mathbf{r}$ (9.15).

The optimal matrix $\mathbf{W}$ that minimizes the empirical risk (9.16) can be obtained via the EVD (9.2) of the sample covariance matrix $\mathbf{Q}$ (9.9). Note that we have used the EVD of $\mathbf{Q}$ already for PCA in Sect. 9.2 (see (9.8)). Remember that PCA uses the n eigenvectors $\mathbf{u}^{(1)}, \dots, \mathbf{u}^{(n)}$ corresponding to the n largest eigenvalues of $\mathbf{Q}$. In contrast, to minimize (9.16), we need to use a different set of eigenvectors in the rows of $\mathbf{W}$ in general. To find the right set of n eigenvectors, we need the sample cross-correlation vector

$$\mathbf{q} := (1/m) \sum_{i=1}^{m} y^{(i)} \mathbf{z}^{(i)}. \tag{9.17}$$

The entry q_j of the vector $\mathbf{q}$ estimates the correlation between the raw feature $z_j^{(i)}$ and the label $y^{(i)}$. We then define the index set

$$\mathcal{S} := \{j_1, \dots, j_n\} \text{ such that } (q_j)^2/\lambda_j \geq (q_{j'})^2/\lambda_{j'} \text{ for any } j \in \mathcal{S}, j' \in \{1, \dots, d\} \notin \mathcal{S}. \tag{9.18}$$

It can then be shown that the rows of the optimal compression matrix $\mathbf{W}$ are the eigenvectors $\mathbf{u}^{(j)}$ with $j \in \mathcal{S}$. We summarize the overall feature learning method in Algorithm 17.

Algorithm 17 Linear Feature Learning for Labeled Data

Input: dataset $\left(\mathbf{z}^{(1)}, y^{(1)}\right), \dots, \left(\mathbf{z}^{(m)}, y^{(m)}\right)$ with raw features $\mathbf{z}^{(i)} \in \mathbb{R}^d$ and numeric labels $y^{(i)} \in \mathbb{R}$; number n of new features.
1: compute EVD (9.10) of the sample covariance matrix (9.9) to obtain orthonormal eigenvectors $\left(\mathbf{u}^{(1)}, \dots, \mathbf{u}^{(d)}\right)$ corresponding to (decreasingly ordered) eigenvalues $\lambda^{(1)} \geq \lambda^{(2)} \geq \dots \geq \lambda^{(d)} \geq 0$
2: compute the sample cross-correlation vector (9.17) and, in turn, the sequence

$$\left(q_1\right)^2/\lambda_1, \dots, \left(q_d\right)^2/\lambda_d \tag{9.19}$$

3: determine indices $i_1, \dots, i_n$ of n largest elements in (9.19)
4: construct compression matrix $\mathbf{W} := \left(\mathbf{u}^{(i_1)}, \dots, \mathbf{u}^{(i_n)}\right)^T \in \mathbb{R}^{n \times d}$
5: compute feature vector $\mathbf{x}^{(i)} = \mathbf{W} \mathbf{z}^{(i)}$
Output: $\mathbf{x}^{(i)}$, for $i = 1, \dots, m$, and compression matrix $\mathbf{W}$.

The main focus of this section was on regression problems involving data points with numeric labels. Given the raw features and labels of the data point in the dataset

$\mathcal{D}$, Algorithm 17 determines new feature vectors $\mathbf{x}^{(i)}$ that allow to linearly predict a numeric label with minimum squared error. A similar approach can be used for classification problems involving data points with discrete labels. The resulting linear feature learning methods are known as **linear discriminant analysis** or **Fisher discriminant analysis** [3].

9.5 Privacy-Preserving Feature Learning

Many important application domains of ML involve sensitive data that is subject to data protection law [9]. Consider a health-care provider (such as a hospital) holding a large database of patient records. From a ML perspective this databases is nothing but a (typically large) set of data points representing individual patients. The data points are characterized by many features including personal identifiers (name, social security number), bio-physical parameters as well as examination results. We could apply ML to learn a predictor for the risk of particular disease given the features of a data point.

Given large patient databases, the ML methods might not be implemented locally at the hospital but using cloud computing. However, data protection requirements might prohibit the transfer of raw patient records that allow to match individuals with bio-physical properties. In this case we might apply feature learning methods to construct new features for each patient such that they allow to learn an accurate hypothesis for predicting a disease but do not allow to identify sensitive properties of the patient such as its name or a social security number.

Let us formalize the above application by characterizing each data point (patient in the hospital database) using raw feature vector $\mathbf{z}^{(i)} \in \mathbb{R}^d$ and a sensitive numeric property $\pi^{(i)}$. We would like to find a compression map $\mathbf{W}$ such that the resulting features $\mathbf{x}^{(i)} = \mathbf{W}\mathbf{z}^{(i)}$ do not allow to accurately predict the sensitive property $\pi^{(i)}$. The prediction of the sensitive property is restricted to be a linear $\hat{\pi}^{(i)} := \mathbf{r}^T \mathbf{x}^{(i)}$ with some weight vector $\mathbf{r}$.

Similar to Sect. 9.4 we want to find a compression matrix $\mathbf{W}$ that transforms, in a linear fashion, the raw feature vector $\mathbf{z} \in \mathbb{R}^d$ to a new feature vector $\mathbf{x} \in \mathbb{R}^n$. However the design criterion for the optimal compression matrix $\mathbf{W}$ was different in Sect. 9.4 where the new feature vectors should allow for an accurate linear prediction of the label. In contrast, here we want to construct feature vectors such that there is no accurate linear predictor of the sensitive property $\pi^{(i)}$.

As in Sect. 9.4, the optimal compression matrix $\mathbf{W}$ is given row-wise by the eigenvectors of the sample covariance matrix (9.9). However, the choice of which eigenvectors to use is different and based on the entries of the sample cross-correlation vector

$$\mathbf{c} := (1/m) \sum_{i=1}^{m} \pi^{(i)} \mathbf{z}^{(i)}. \tag{9.20}$$

We summarize the construction of the optimal privacy-preserving compression matrix and corresponding new feature vectors in Algorithm 18.

Algorithm 18 Privacy Preserving Feature Learning

Input: dataset $(\mathbf{z}^{(1)}, y^{(1)}), \ldots, (\mathbf{z}^{(m)}, y^{(m)})$ with raw features $\mathbf{z}^{(i)} \in \mathbb{R}^d$ and (numeric) sensitive property $\pi^{(i)} \in \mathbb{R}$; number n of new features.

1: compute the EVD (9.10) of the sample-covariance matrix (9.9) to obtain orthonormal eigenvectors $(\mathbf{u}^{(1)}, \ldots, \mathbf{u}^{(d)})$ corresponding to (decreasingly ordered) eigenvalues $\lambda_1 \geq \lambda_2 \geq \ldots \geq \lambda_d \geq 0$

2: compute the sample cross-correlation vector (9.20) and, in turn, the sequence

$$(c_1)^2/\lambda_1, \ldots, (c_d)^2/\lambda_d \qquad (9.21)$$

3: determine indices $i_1, \ldots, i_n$ of n smallest elements in (9.21)

4: construct compression matrix $\mathbf{W} := \left(\mathbf{u}^{(i_1)}, \ldots, \mathbf{u}^{(i_n)}\right)^T \in \mathbb{R}^{n \times d}$

5: compute feature vector $\mathbf{x}^{(i)} = \mathbf{W}\mathbf{z}^{(i)}$

Output: privacy-preserving feature vectors $\mathbf{x}^{(i)}$, for $i = 1, \ldots, m$, and compression matrix $\mathbf{W}$.

Algorithm 18 learns a map $\mathbf{W}$ to extract privacy-preserving features out of the raw feature vector of a data point. These new features are privacy-preserving as they do not allow to accurately predict (in a linear fashion) a sensitive property π of the data point. Another formalization for the preservation of privacy can be obtained using information-theoretic concepts. This information-theoretic approach interprets data points, their feature vector and sensitive property, as realizations of RVs. It is then possible to use the mutual information between new features $\mathbf{x}$ and the sensitive (private) property π as an optimization criterion for learning a compression map h (Sect. 9.1). The resulting feature learning method (referred to as privacy-funnel) differs from Algorithm 18 not only in the optimization criterion for the compression map but also in that it allows it to be non-linear [10, 11].

9.6   Random Projections

Note that PCA involves an EVD of the sample covariance matrix $\mathbf{Q}$ (9.9). The computational complexity (e.g., measured by number of multiplications and additions) for computing this EVD is lower bounded by $\min\{D^2, m^2\}$ [12, 13]. This computational complexity can be prohibitive for ML applications with n and m being of the order of millions (which is already the case if the features are pixel values of a 512×512 RGB bitmap, see Sect. 2.1.1).

There is a computationally cheap alternative to PCA (Algorithm 15) for finding a useful compression matrix $\mathbf{W}$ in (9.5). This alternative is to construct the compression matrix $\mathbf{W}$ entry-wise

$$W_{i,j} := a^{(i,j)} \text{ with i.i.d. } a_{i,j} \sim p(a). \qquad (9.22)$$

The entries of the matrix (9.22) are realizations of i.i.d. RVs $a_{i,j}$ with some common probability distribution $p(a)$. Different choices for the probability distribution $p(a)$ have been studied in the literature [14]. The Bernoulli distribution is used to obtain a compression matrix with binary entries. Another popular choice for $p(a)$ is the multivariate normal (Gaussian) distribution.

Consider data points whose raw feature vectors $\mathbf{z}$ are located near a s-dimensional subspace of $\mathbb{R}^d$. The feature vectors $\mathbf{x}$ obtained via (9.5) using a random matrix (9.22) allows to reconstruct the raw feature vectors $\mathbf{z}$ with high probability whenever

$$n \geq Cs \log d. \tag{9.23}$$

The constant C depends on the maximum tolerated reconstruction error η (such that $\|\widehat{\mathbf{z}} - \mathbf{z}\|_2^2 \leq \eta$ for any data point) and the probability that the features $\mathbf{x}$ (see)(9.22)) allow for a maximum reconstruction error η [14, Theorem 9.27.].

9.7 Dimensionality Increase

The focus of this chapter is on dimensionality reduction methods that learn a feature map delivering new feature vectors which are (significantly) shorter than the raw feature vectors. However, it might sometimes be beneficial to learn a feature map that delivers new feature vectors which are longer than the raw feature vectors. We have already discussed two examples for such feature learning methods in Sects. 3.2 and 3.9. Polynomial regression maps a single raw feature z to a feature vector containing the powers of the raw feature z. This allows to use apply linear predictor maps to the new feature vectors to obtain predictions that depend non-linearly on the raw feature z. Kernel methods might even use a feature map that delivers feature vectors belonging to an infinite-dimensional Hilbert space [15].

Mapping raw feature vectors into higher-dimensional (or even infinite-dimensional) spaces might be useful if the intrinsic geometry of the data points is simpler when looked at in the higher-dimensional space. Consider a binary classification problem where data points are highly inter-winded in the original feature space (see Fig. 3.7). Loosely speaking, mapping into higher-dimensional feature space might "flatten-out" a non-linear decision boundary between data points. We can then apply linear classifiers to the higher-dimensional features to achieve accurate predictions.

9.8 Exercises

Exercise 9.1 Computational Burden of Many Features Discuss the computational complexity of linear regression. How much computation do we need to compute the linear predictor that minimizes the average squared error on a training set?

Exercise 9.2 Power Iteration The key computational step of PCA amounts to an EVD of the psd matrix (9.9). Consider an arbitrary initial vector $\mathbf{u}^{(r)}$ and the sequence obtained by iterating

$$\mathbf{u}^{(r+1)} := \mathbf{Q}\mathbf{u}^{(r)}/\left\| \mathbf{Q}\mathbf{u}^{(r)} \right\|. \tag{9.24}$$

Under what (if any) conditions for the initialization $\mathbf{u}^{(r)}$ can be ensure that the sequence $\mathbf{u}^{(r)}$ converges to the eigenvector $\mathbf{u}^{(1)}$ of $\mathbf{Q}$ corresponding to its largest eigenvalue λ_1

Exercise 9.3 Linear Classifiers with High-Dimensional Features Consider a training set $\mathcal{D}$ consisting of $m = 10^{10}$ labeled data points $\left(\mathbf{z}^{(1)}, y^{(1)}\right), \ldots, \left(\mathbf{z}^{(m)}, y^{(m)}\right)$ with raw feature vectors $\mathbf{z}^{(i)} \in \mathbb{R}^{4000}$ and binary labels $y^{(i)} \in \{-1, 1\}$. Assume we have used a feature learning method to obtain the new features $\mathbf{x}^{(i)} \in \{0, 1\}^n$ with $n = m$ and such that the only non-zero entry of $\mathbf{x}^{(i)}$ is $x_i^{(i)} = 1$, for $i = 1, \ldots, m$. Can you find a linear classifier that perfectly classifies the training set?

References

1. I. Goodfellow, Y. Bengio, A. Courville, *Deep Learning* (MIT Press, Cambridge, 2016)
2. G. Strang, *Computational Science and Engineering* (Wellesley-Cambridge Press, MA, 2007)
3. T. Hastie, R. Tibshirani, J. Friedman, *The Elements of Statistical Learning*. Springer Series in Statistics (Springer, New York, 2001)
4. J. Wright, Y. Peng, Y. Ma, A. Ganesh, S. Rao, Robust principal component analysis: exact recovery of corrupted low-rank matrices by convex optimization, in *Neural Information Processing Systems, NIPS 2009* (2009)
5. S. Roweis, EM Algorithms for PCA and SPCA. *Advances in Neural Information Processing Systems* (MIT Press, Cambridge, 1998), pp. 626–632
6. M.E. Tipping, C. Bishop, Probabilistic principal component analysis. J. Roy. Stat. Soc. B **21**(3), 611–622 (1999)
7. M.E.J. Newman, *Networks: An Introduction* (Oxford University Press, Oxford, 2010)
8. U. von Luxburg, A tutorial on spectral clustering. Stat. Comput. **17**(4), 395–416 (2007)
9. S. Wachter, Data protection in the age of big data. Nat. Electron. **2**(1), 6–7 (2019)
10. A. Makhdoumi, S. Salamatian, N. Fawaz, M. Médard, From the information bottleneck to the privacy funnel, in *2014 IEEE Information Theory Workshop (ITW 2014)*, pp. 501–505 (2014)
11. Y.Y. Shkel, R.S. Blum, H.V. Poor, Secrecy by design with applications to privacy and compression. IEEE Trans. Inf. Theory **67**(2), 824–843 (2021)
12. Q. Du, J. Fowler, Low-complexity principal component analysis for hyperspectral image compression. Int. J. High Perform. Comput. Appl., pp. 438–448 (2008)
13. A. Sharma, K. Paliwal, Fast principal component analysis using fixed-point analysis. Pattern Recogn. Lett. **28**, 1151–1155 (2007)
14. S. Foucart, H. Rauhut, *A Mathematical Introduction to Compressive Sensing* (Springer, New York, 2012)
15. B. Schölkopf, A. Smola, *Learning with Kernels: Support Vector Machines, Regularization, Optimization, and Beyond* (MIT Press, Cambridge, 2002)

Chapter 10
Transparent and Explainable ML

The successful deployment of ML methods depends on their transparency (or explainability). We refer to techniques that aim at making ML method transparent (or explainable) as explainable ML. Providing explanations for the predictions of a ML method is particularly important when these predictions inform decision making [1]. Explanations for automated decision making system have become a legal requirement [2].

Even for applications where predictions are not directly used to inform far-reaching decisions, providing explanations is important. The human end users have an intrinsic desire for explanations that resolve the uncertainty about the prediction. This is known as the "need for closure" in psychology [3, 4]. Beside legal and psychological requirements, providing explanations for predictions might also be useful for validating and verifying ML methods. Indeed, the explanations provided for preditions might point the user (domain expert) to incorrect modelling assumptions underlying the ML method.

Explainable ML is challenging since explanations must be tailored (personalized) to human end-users with varying backgrounds and in different contexts [5]. The user background includes the formal education as well as the individual digital literacy. Some users might have received university-level education in ML, while other users might have no relevant formal training (such as an undergraduate course in linear algebra). Linear regression with few features might be perfectly interpretable for the first group but be considered a "black box" for the latter. To enable tailored explanations we need to model the user background as relevant for understanding the ML predictions.

A. Jung, *Machine Learning*, Machine Learning: Foundations, Methodologies, and Applications, https://doi.org/10.1007/978-981-16-8193-6_10

This chapter discusses explainable ML methods that have access to some user signal or feedback for some data points. Such a user signal might be obtained in various ways, including answers to surveys or bio-physical measurements collected via wearables or medical diagnostics. The user signal is used to determine (to some extent) the end-user background and, in turn, to tailor the delivered explanations for this end-user.

Existing explainable ML methods can be roughly divided into two categories. The first category is referred to as "model agnostic" (or "black box explainable ML" [1]). Model agnostic methods do not require knowledge of the detailed work principles of a ML method. In particular, these methods do not require knowledge of the hypothesis space used by a ML method. They earn how to explain the predictions of a ML method by observing its predictions for a set of data points [6]. A second category of explainable ML methods (referred to as "white-box" methods [1])uses ML methods that are considered as intrinsically explainable. The intrinsic explainability of a ML method depends crucially on its choice for the hypothesis space (see Sect. 2.2). This chapter discusses one recent method from each of the two explainable ML categories [7, 8]. The common theme of both methods is the use of information-theoretic concepts to measure the usefulness of explanations [9].

Section 10.1 discusses a recently proposed model-agnostic approach to explainable ML that constructs tailored explanations for the predictions of a given ML method [7]. This approach does not require any details about the internal mechanism of a ML method whose predictions are to be explained. Rather, this approach only requires a (sufficiently large) training set of data points for which the predictions of the ML method are known. To tailor the explanations to a particular user, we use the values of a user (feedback) signal provided for the data points in the training set. Roughly speaking, the explanations are chosen such that they maximally reduce the "surprise" or uncertainty that the user has about the predictions of the ML method.

Section 10.2 discusses an example for a ML method that uses a hypothesis space that is intrinsically explainable [8]. This explainable hypothesis space is obtained by pruning an arbitrary hypothesis space such as linear maps (see Sect. 3.1) or non-linear maps represented by either a deep neural network (see Sect. 3.11) or decision trees (see Sect. 3.10). This pruning is implemented via adding a regularization term to ERM (4.3), resulting in an instance of SRM (7.2) which we refer to as empirical risk minimization (EERM). The regularization term favours hypotheses that are explainable to a user. Similar to the method in Sect. 10.1, the explainability of a map is quantified by information theoretic quantities. For example, if the original hypothesis space is the set of linear maps using a large number of features, the regularization term might favour maps that depend only on few features that are interpretable. Hence, we can interpret EERM as a feature learning method that aims at learning relevant and interpretable features (see Chapter 9).

10.1 A Model Agnostic Method

Consider a ML application involving data points with features $\mathbf{x} = (x_1, \ldots, x_n)^T \in \mathbb{R}^n$ and label $y \in \mathbb{R}$ and some ML method that, given some labelled training data

$$(\mathbf{x}^{(1)}, y^{(1)}), (\mathbf{x}^{(2)}, y^{(2)}), \ldots, (\mathbf{x}^{(m)}, y^{(m)}), \tag{10.1}$$

learn a predictor (map)

$$h(\cdot) : \mathbb{R}^n \to \mathbb{R} : \mathbf{x} \mapsto \hat{y} = h(\mathbf{x}). \tag{10.2}$$

The precise working principle of this ML method for how to learn this hypothesis h is not relevant in what follows.

The learnt predictor $h(\mathbf{x})$ is applied to the features of a data point to obtain the predicted label $\hat{y} := h(\mathbf{x})$. The prediction $\hat{y}$ is then delivered to a human end-user. Depending on the ML application, this end-user might be a streaming service subscriber [10], a dermatologist [11] or a city planner [12].

Human users of ML methods often have some conception or model for the relation between features $\mathbf{x}$ and label y of a data point. This intrinsic model might vary significantly between users with different (social or educational) background. We will model the user understanding of a data point by a "user summary" $u \in \mathbb{R}$. The summary is obtained by a (possibly stochastic) map from the features $\mathbf{x}$ of a data point. For ease of exposition, we will focus on summaries obtained by a deterministic map

$$u(\cdot) : \mathbb{R}^n \to \mathbb{R} : \mathbf{x} \mapsto u := u(\mathbf{x}). \tag{10.3}$$

However, our approach also covers stochastic maps characterized by a conditional probability distribution $p(u|\mathbf{x})$.

The (user-specific) quantity u is determined by the features $\mathbf{x}$ of a data point. We might think of the value u for a specific data point as a signal that reflects how the human end-user interprets (or perceives) the data point, given her knowledge (including formal education) and the context of the ML application. We do not assume any knowledge about the details for how the signal value u is formed for a specific data point. In particular, we do not know any properties of the map $u(\cdot) : \mathbf{x} \mapsto u$.

Our approach is quite flexible as it allows for very different forms of user summaries. The user summary could be the prediction obtained from a simplified model, such as linear regression using few features that the user anticipates as being relevant. Another example for a user summary u could be a higher-level feature, such as eye spacing in facial pictures, that the user considers relevant [13].

Note that, since we allow for an arbitrary map in (10.3), the user summary $u(\mathbf{x})$ obtained for a random data point with features $\mathbf{x}$ might be correlated with the prediction $\hat{y} = h(\mathbf{x})$. As an extreme case, consider highly knowledgable users that are able to predict the labels of data points form their features as well as the ML method. In

Fig. 10.1 An explanation e
provides additional
information $I(\hat{y}, e|u)$ to a
user u about the prediction $\hat{y}$

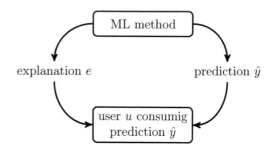

this case, the maps (10.2) and (10.3) might be nearly identical. However, in general
the predictions delivered by the learnt hypothesis (10.2) will be different from the
user summary.

We formalize the act of explaining a prediction $\hat{y} = h(\mathbf{x})$ as presenting some
additional quantity e to the user (see Figure 10.1). This explanation e can be any
artefact that helps the user to understand the prediction $\hat{y}$, given her understanding u
of the data point. Loosely speaking, the aim of providing explanation e is to reduce
the uncertainty of the user u about the prediction $\hat{y}$ [4].

Our approach is quite flexible in that it allows for many different forms of explanations. An explanation could be a subset of features of a data point (see [14] and Sect.
10.1.2). More generally, explanations could be obtained from simple local statistics
(averages) of features that are considered related, such as near-by pixels in an image
or consecutive samples of an audio signal. Instead of individual features, carefully
chosen data points can also serve as an explanation [15, 16].

For the sake of exposition, our focus will be on explanations obtained via a deterministic map

$$e(\cdot) : \mathbb{R}^n \to \mathbb{R} : \mathbf{x} \mapsto e := e(\mathbf{x}), \qquad (10.4)$$

from the features $\mathbf{x}$ of a data point. However, our approach can be generalized without
difficulty to handle explanations obtained by a (stochastic) map. In the end, we only
require the specification of the conditional probability distribution $p(e|\mathbf{x})$.

The explanation e (10.4) depends only on the features $\mathbf{x}$ but not explicitly on
the prediction $\hat{y}$. However, our method for constructing the map (10.4) takes into
account the properties of the predictor map $h(\mathbf{x})$ (10.2). In particular, Algorithm 19
below requires as input the predicted labels $\hat{y}^{(i)}$ for a set of data points (that serve as
a training set for our method).

To obtain comprehensible explanations that can be computed efficiently, we must
typically restrict the space of possible explanations to a small subset $\mathcal{F}$ of maps
(10.4). This is conceptually similar to the restriction of the space of possible predictor
functions in a ML method to a small subset of maps which is known as the hypothesis
space.

10.1.1 Probabilistic Data Model for XML

In what follows, we model data points as realizations of i.i.d. RVs with common (joint) probability distribution $p(\mathbf{x}, y)$ of features and label (see Sect. 2.1.4). Modelling the data points as realizations of RVs implies that the user summary u, prediction $\hat{y}$ and explanation e are also RVs. The joint distribution $p(u, \hat{y}, e, \mathbf{x}, y)$ conforms with the Bayesian network [17] depicted in Figure 10.2. Indeed,

$$p(u, \hat{y}, e, \mathbf{x}, y) = p(u|\mathbf{x}) \cdot p(e|\mathbf{x}) \cdot p(\hat{y}|\mathbf{x}) \cdot p(\mathbf{x}, y). \qquad (10.5)$$

We measure the amount of additional information provided by an explanation e for a prediction $\hat{y}$ to some user u via the conditional mutual information (MI) [9, Ch. 2 and 8]

$$I(e; \hat{y}|u) := \mathbb{E}\left\{ \log \frac{p(\hat{y}, e|u)}{p(\hat{y}|u)p(e|u)} \right\}. \qquad (10.6)$$

The conditional MI $I(e; \hat{y}|u)$ can also be interpreted as a measure for the amount by which the explanation e reduces the uncertainty about the prediction $\hat{y}$ which is delivered to some user u. Providing the explanation e serves the apparent human need to understand observed phenomena, such as the predictions from a ML method [4].

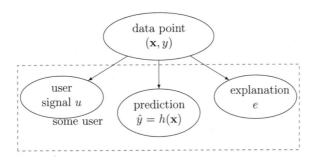

Fig. 10.2 A simple probabilistic graphical model (a Bayesian network [18, 19]) for explainable ML. We interpret data points (with features $\mathbf{x}$ and label y) along with the user summary u, e and predicted label $\hat{y}$ as realizations of RVs. These RVs satisfy conditional independence relations encoded by the directed links of the graph [19]. In particular, given the data point, the predicted label $\hat{y}$ obtained from a "black box" ML method, the explanation e obtained from our method and the user summary u are conditionally independent. This conditional independence is trivial if all these quantities are obtained from deterministic maps applied to the features $\mathbf{x}$ of the data point

10.1.2   Computing Optimal Explanations

Capturing the effect of an explanation using the probabilistic model (10.6) offers a principled approach to computing an optimal explanation e. We require the optimal explanation e^* to maximize the conditional MI (10.6) between the explanation e and the prediction $\hat{y}$ conditioned on the user summary u of the data point.

Formally, an optimal explanation e^* solves

$$I(e^*; \hat{y}|u) = \sup_{e \in \mathcal{F}} I(e; \hat{y}|u). \tag{10.7}$$

The choice for the subset $\mathcal{F}$ of valid explanations offers a trade-off between comprehensibility, informativeness and computational cost incurred by an explanation e^* (solving (10.7)).

The maximization problem (10.7) for obtaining optimal explanations is similar to the approach in [6]. While [6] uses the unconditional MI between explanation and prediction, (10.7) uses the conditional MI given the user summary u. Thus, our approach provides personalized explanations that are tailored to the user, as characterized by her summary u.

Let us illustrate the concept of optimal explanations (10.7) using a linear regression method. We model the features $\mathbf{x}$ as a realization of a multivariate normal random vector with zero mean and covariance matrix $\mathbf{C}_x$,

$$\mathbf{x} \sim \mathcal{N}(\mathbf{0}, \mathbf{C}_x). \tag{10.8}$$

The predictor and the user summary are linear functions

$$\hat{y} := \mathbf{w}^T \mathbf{x}, \text{ and } u := \mathbf{v}^T \mathbf{x}. \tag{10.9}$$

We construct explanations via subsets of individual features x_j that are considered most relevant for a user to understand the prediction $\hat{y}$ (see [20, Definition 2] and [21]). Thus, we consider explanations of the form

$$e := \{x_i\}_{i \in \mathcal{E}} \text{ with some subset } \mathcal{E} \subseteq \{1, \ldots, n\}. \tag{10.10}$$

The complexity of an explanation e is measured by the number $|\mathcal{E}|$ of features that contribute to it. We limit the complexity of explanations by a fixed (small) sparsity level,

$$|\mathcal{E}| \leq s(\ll n). \tag{10.11}$$

Modelling the feature vector $\mathbf{x}$ as Gaussian (10.8) implies that the prediction $\hat{y}$ and user summary u obtained from (10.9) is jointly Gaussian for a given $\mathcal{E}$ (10.4). Basic properties of multivariate normal distributions [9, Ch. 8], allow to develop (10.7) as

$$\max_{\substack{\mathcal{E}\subseteq\{1,\ldots,n\} \\ |\mathcal{E}|\le s}} I(e;\hat{y}|u)$$

$$= h(\hat{y}|u) - h(\hat{y}|u,\mathcal{E})$$
$$= (1/2)\log \mathbf{C}_{\hat{y}|u} - (1/2)\log\det \mathbf{C}_{\hat{y}|u,\mathcal{D}^{(\text{train})}}$$
$$= (1/2)\log \sigma^2_{\hat{y}|u} - (1/2)\log \sigma^2_{\hat{y}|u,\mathcal{D}^{(\text{train})}}. \tag{10.12}$$

Here, $\sigma^2_{\hat{y}|u}$ denotes the conditional variance of the prediction $\hat{y}$, conditioned on the user summary u. Similarly, $\sigma^2_{\hat{y}|u,\mathcal{E}}$ denotes the conditional variance of $\hat{y}$, conditioned on the user summary u and the subset $\{x_j\}_{j\in\mathcal{E}}$ of features. The last step in (10.12) follows from the fact that $\hat{y}$ is a scalar random variable.

The first component of the final expression of (10.12) does not depend on the index set $\mathcal{E}$ used to construct the explanation e (see (10.10)). Therefore, the optimal choice for $\mathcal{E}$ solves

$$\sup_{|\mathcal{E}|\le s} -(1/2)\log \sigma^2_{\hat{y}|u,\mathcal{E}}. \tag{10.13}$$

The maximization (10.13) is equivalent to

$$\inf_{|\mathcal{E}|\le s} \sigma^2_{\hat{y}|u,\mathcal{E}}. \tag{10.14}$$

In order to solve (10.14), we relate the conditional variance $\sigma^2_{\hat{y}|u,\mathcal{E}}$ to a particular decomposition

$$\hat{y} = \eta u + \sum_{j\in\mathcal{E}} \beta_j x_j + \varepsilon. \tag{10.15}$$

For an optimal choice of the coefficients η and β_j, the variance of the error term in (10.15) is given by $\sigma^2_{\hat{y}|u,\mathcal{E}}$. Indeed,

$$\min_{\eta,\beta_j\in\mathbb{R}} \mathbb{E}\left\{\left(\hat{y} - \eta u - \sum_{j\in\mathcal{E}} \beta_j x_j\right)^2\right\} = \sigma^2_{\hat{y}|u,e}. \tag{10.16}$$

Inserting (10.29) into (10.14), an optimal choice $\mathcal{E}$ (of feature) for the explanation of prediction $\hat{y}$ to user u is obtained from

$$\inf_{|\mathcal{E}|\le s} \min_{\eta,\beta_j\in\mathbb{R}} \mathbb{E}\left\{\left(\hat{y} - \eta u - \sum_{j\in\mathcal{E}} \beta_j x_j\right)^2\right\} \tag{10.17}$$

$$= \min_{\|\beta\|_0\le s} \mathbb{E}\left\{\left(\hat{y} - \eta u - \beta^T\mathbf{x}\right)^2\right\}. \tag{10.18}$$

An optimal subset $\mathcal{E}_{\text{opt}}$ of features defining the explanation e (10.10) is obtained from any solution β_{opt} of (10.18) via

$$\mathcal{E}_{\text{opt}} = \operatorname{supp}\beta_{\text{opt}}. \tag{10.19}$$

Under a Gaussian model (10.8), Sect. 10.1.2 shows how to construct optimal explanations via the (support of the) solutions β_{opt} of the sparse linear regression problem (10.18). To obtain a practical algorithm for computing (approximately) optimal explanations (10.19), we approximate the expectation in (10.18) with an average over the training data points $(\mathbf{x}^{(i)}, \hat{y}^{(i)}, u^{(i)})$. This resulting method for computing personalized explanations is summarized in Algorithm 19.

Algorithm 19 XML Algorithm

Input: explanation complexity s, training samples $(\mathbf{x}^{(i)}, \hat{y}^{(i)}, u^{(i)})$ for $i = 1, \ldots, m$
1: compute $\widehat{\beta}$ by solving

$$\widehat{\beta} \in \operatorname*{argmin}_{\eta \in \mathbb{R}, \|\beta\|_0 \leq s} \sum_{i=1}^{m} \left(\hat{y}^{(i)} - \eta u^{(i)} - \beta^T \mathbf{x}^{(i)} \right)^2 \tag{10.20}$$

Output: feature set $\widehat{\mathcal{E}} := \operatorname{supp} \widehat{\beta}$

Note that Algorithm 19 is interactive since the user has to provide samples $u^{(i)}$ of its summary for the data points with features $\mathbf{x}^{(i)}$. Based on the user input $u^{(i)}$, for $i = 1, \ldots, m$, Algorithm 19 learns an optimal subset $\mathcal{E}$ of features (10.10) that are used for the explanation of predictions.

The sparse regression problem (10.20) becomes intractable for large feature length n. However, if the features are weakly correlated with each other and the user summary u, the solutions of (10.20) can be found by efficient convex optimization methods. Indeed, for a wide range of settings, sparse regression (10.20) can be solved via a convex relaxation, known as the least absolute shrinkage and selection operator (Lasso) [22],

$$\widehat{\beta} \in \operatorname*{argmin}_{\eta \in \mathbb{R}, \beta \in \mathbb{R}^n} \sum_{i=1}^{m} \left(\hat{y}^{(i)} - \eta u^{(i)} - \beta^T \mathbf{x}^{(i)} \right)^2 + \lambda \|\beta\|_1. \tag{10.21}$$

We have already a good understanding of choosing the Lasso parameter λ in (10.21) such that its solutions coincide with the solutions of (10.20) (see, e.g., [22]).

10.2   Explainable Empirical Risk Minimization

Section 7.1 discussed SRM (7.1) as a method for pruning the hypothesis space $\mathcal{H}$ used in ERM (4.3). This pruning is implemented either via a (hard) constraint as in (7.1) or by adding a regularization term to the training error as in (7.2). The idea of SRM is to avoid (prune away) hypothesis maps that perform good on the training set but poorly outside (they do not generalize). Here, we will use another criterion for steering the pruning and construction of regularization terms. In particular, we use the (intrinsic) explainability of a hypotheses maps as a regularization term.

To make the notion of explainability precise we will use the probabilistic model of Sect. 10.1.1. We interpret data points as realizations of i.i.d.. RVs with common (joint) probability distribution $p(\mathbf{x}, y)$ of features $\mathbf{x}$ and label y. A quantitative measure the intrinsic explainability of a hypothesis $h \in \mathcal{H}$ is the conditional (differential) entropy [9, Ch. 2 and 8]

$$H(\hat{y}|u) := -\mathbb{E}\left\{ \log p(\hat{y}|u) \right\}. \tag{10.22}$$

The conditional entropy (10.22) indicates the uncertainty about the prediction $\hat{y}$, given the user summary $\hat{u} = u(\mathbf{x})$. Smaller values $H(\hat{y}; u)$ correspond to smaller levels of uncertainty in the predictions $\hat{y}$ that is experienced by user u.

We obtain Explainable empirical risk minimization (explainable empirical risk minimization) by requiring a sufficiently small conditional entropy (10.22) of a hypothesis,

$$\hat{h} \in \underset{h \in \mathcal{H}}{\arg\min}\, \widehat{L}(h) \quad \text{s.t.} \quad H(\hat{y}|\hat{u}) \leq \eta. \tag{10.23}$$

The random variable $\hat{y} = h(\mathbf{x})$ in the constraint of (10.23) is obtained by applying the predictor map $h \in \mathcal{H}$ to the features. The constraint $H(\hat{y}|\hat{u}) \leq \eta$ in (10.23) enforces the learnt hypothesis $\hat{h}$ to be sufficiently explainable in the sense that the conditional entropy $H(\hat{h}|\hat{u}) \leq \eta$ does not exceed a prescribed level η.

Let us now consider the special case of EERM (10.23) for the linear hypothesis space

$$h^{(\mathbf{w})}(\mathbf{x}) := \mathbf{w}^T \mathbf{x} \text{ with some weight vector } \mathbf{w} \in \mathbb{R}^n. \tag{10.24}$$

Moreover, we assume that the features $\mathbf{x}$ of a data point and its user summary u are jointly Gaussian with mean zero and covariance matrix $\mathbf{C}$,

$$\left(\mathbf{x}^T, \hat{u}\right)^T \sim \mathcal{N}(\mathbf{0}, \mathbf{C}). \tag{10.25}$$

Under the assumptions (10.24) and (10.25) (see [9, Ch. 8]),

$$H(\hat{u}|\hat{y}) = (1/2) \log \sigma^2_{\hat{y}|\hat{u}}. \tag{10.26}$$

Here, we used the conditional variance $\sigma^2_{\hat{y}|\hat{u}}$ of $\hat{y}$ given the random user summary $u = u(\mathbf{x})$. Inserting (10.26) into the generic form of EERM (10.23),

$$\hat{h} \in \underset{h \in \mathcal{H}}{\arg\min}\, \widehat{L}(h) \quad \text{s.t.} \quad \log \sigma^2_{\hat{y}|\hat{u}} \leq \eta. \tag{10.27}$$

By the monotonicity of the logarithm, (10.27) is equivalent to

$$\hat{h} \in \underset{h \in \mathcal{H}}{\operatorname{argmin}} \widehat{L}(h) \quad \text{s.t.} \quad \sigma_{\hat{y}|\hat{u}}^2 \le e^{(\eta)}. \tag{10.28}$$

To further develop (10.16), we use the identity

$$\min_{\eta \in \mathbb{R}} \mathbb{E}\{(\hat{y} - \eta u)^2\} = \sigma_{\hat{y}|\hat{u}}^2. \tag{10.29}$$

The identity (10.29) relates the conditional variance $\sigma_{\hat{y}|\hat{u}}^2$ to the minimum mean squared error that can be achieved by estimating $\hat{y}$ using a linear estimator $\eta \hat{u}$ with some $\eta \in \mathbb{R}$. Inserting (10.29) and (10.24) into (10.28),

$$\hat{h} \in \underset{\mathbf{w} \in \mathbb{R}^n, \eta \in \mathbb{R}}{\operatorname{argmin}} \widehat{L}(h^{(\mathbf{w})}) \quad \text{s.t.} \quad \mathbb{E}\{ \underbrace{(\mathbf{w}^T \mathbf{x} - \eta \hat{u})^2}_{\overset{(10.24)}{=} \hat{y}} \} \le e^{(\eta)}. \tag{10.30}$$

The inequality constraint in (10.30) is convex [23, Ch. 4.2.]. For squared error loss, the objective function $\widehat{L}(h^{(\mathbf{w})})$ is also convex. Thus, for linear least squares regression, we can reformulate (10.30) as an equivalent (dual) unconstrained problem [23, Ch. 5]

$$\hat{h} \in \underset{\mathbf{w} \in \mathbb{R}^n, \eta \in \mathbb{R}}{\operatorname{argmin}} \mathcal{E}(h^{(\mathbf{w})}) + \lambda \mathbb{E}\{(\mathbf{w}^T \mathbf{x} - \eta \hat{u})^2\}. \tag{10.31}$$

By convex duality, for a given threshold $e^{(\eta)}$ in (10.30), we can find a value for λ in (10.31) such that (10.30) and (10.31) have the same solutions [23, Ch. 5]. Algorithm 20 below is obtained from (10.31) by approximating the expectation $\mathbb{E}\{(\mathbf{w}^T \mathbf{x} - \eta \hat{u})^2\}$ with an average over the training data points $(\mathbf{x}^{(i)}, \hat{y}^{(i)}, \hat{u}^{(i)})$ for $i = 1, \dots, m$.

Algorithm 20 Explainable Linear Least Squares Regression

Input: explainability parameter λ, training examples $(\mathbf{x}^{(i)}, \hat{y}^{(i)}, \hat{u}^{(i)})$ for $i = 1, \dots, m$
1: solve

$$\widehat{\mathbf{w}} \in \underset{\eta \in \mathbb{R}, \mathbf{w} \in \mathbb{R}^n}{\operatorname{argmin}} \sum_{i=1}^m \underbrace{(\hat{y}^{(i)} - \mathbf{w}^T \mathbf{x}^{(i)})^2}_{\text{empirical risk}} + \lambda \underbrace{(\mathbf{w}^T \mathbf{x}^{(i)} - \eta \hat{u}^{(i)})^2}_{\text{explainablity}} \tag{10.32}$$

Output: weights $\widehat{\mathbf{w}}$ of explainable linear hypothesis

---

10.3 Exercises

Exercise 10.1 (Convexity of Explainable Linear Regression) Rewrite the optimization problem (10.32) as an equivalent quadratic optimization problem $\min_{\mathbf{v}} \mathbf{v}^T \mathbf{Q} \mathbf{v} + \mathbf{v}^T \mathbf{q}$. Identify the matrix $\mathbf{Q}$ and the vector $\mathbf{q}$.

References

1. H.-F. Cheng, R. Wang, Z. Zhang, F. O'Connell, T. Gray, F. M. Harper, and H. Zhu. Explaining decision-making algorithms through UI: Strategies to help non-expert stakeholders. In *Proceedings of the 2019 CHI Conference on Human Factors in Computing Systems*, CHI '19, pages 1–12, New York, NY, USA, 2019. Association for Computing Machinery
2. P. Hacker, R. Krestel, S. Grundmann, and F. Naumann. Explainable AI under contract and tort law: legal incentives and technical challenges. *Artificial Intelligence and Law*, 2020
3. T.K. DeBacker, H.M. Crowson, Influences on cognitive engagement: Epistemological beliefs and need for closure. British Journal of Educational Psychology **76**(3), 535–551 (2006)
4. J. Kagan, Motives and development. Journal of Personality and Social Psychology **22**(1), 51–66 (1972)
5. Q. V. Liao, D. Gruen, and S. Miller. Questioning the ai: Informing design practices for explainable ai user experiences. In *Proceedings of the 2020 CHI Conference on Human Factors in Computing Systems*, CHI '20, pages 1–15, New York, NY, USA, 2020. Association for Computing Machinery
6. J. Chen, L. Song, M. Wainwright, and M. Jordan. Learning to explain: An information-theoretic perspective on model interpretation. In *Proc. 35th Int. Conf. on Mach. Learning*, Stockholm, Sweden, 2018
7. A. Jung, P. Nardelli, An information-theoretic approach to personalized explainable machine learning. IEEE Sig. Proc. Lett. **27**, 825–829 (2020)
8. A. Jung. Explainable empiricial risk minimization. *submitted to IEEE Sig. Proc. Letters (preprint:* https://arxiv.org/pdf/2009.01492.pdf*)*, 2020
9. T.M. Cover, J.A. Thomas, *Elements of Information Theory*, 2nd edn. (Wiley, New Jersey, 2006)
10. C. Gomez-Uribe and N. Hunt. The netflix recommender system: Algorithms, business value, and innovation. *Association for Computing Machinery*, 6(4), 2016
11. A. Esteva, B. Kuprel, R. A. Novoa, J. Ko, S. M. Swetter, H. M. Blau, and S. Thrun. Dermatologist-level classification of skin cancer with deep neural networks. *Nature*, 542, 2017
12. X. Yang and Q. Wang. Crowd hybrid model for pedestrian dynamic prediction in a corridor. *IEEE Access*, 7, 2019
13. K. Jeong, J. Choi, and G. Jang. Semi-local structure patterns for robust face detection. *IEEE Sig. Proc. Letters*, 22(9), 2015
14. M. Ribeiro, S. Singh, and C. Guestrin. "Why should i trust you?": Explaining the predictions of any classifier. In *Proc. 22nd ACM SIGKDD*, pages 1135–1144, 2016
15. J. McInerney, B. Lacker, S. Hansen, K. Higley, H. Bouchard, A. Gruson, and R. Mehrotra. Explore, exploit, and explain: personalizing explainable recommendations with bandits. In *Proceedings of the 12th ACM Conference on Recommender Systems*, 2018
16. M. Ribeiro, S. Singh, and C. Guestrin. Anchors: High-precision model-agnostic explanations. In *Proc. AAAI Conference on Artificial Intelligence (AAAI)*, 2018
17. J. Pearl. *Probabilistic Reasoning in Intelligent Systems*. Morgan Kaufmann, 1988
18. S.L. Lauritzen, *Graphical Models* (Clarendon Press, Oxford, UK, 1996)

19. D. Koller, N. Friedman, *Probabilistic Graphical Models: Principles and Techniques* (MIT Press, Adaptive Computation and Machine Learning, 2009)
20. G. Montavon, W. Samek, K. Müller, Methods for interpreting and understanding deep neural networks. Digital Signal Processing **73**, 1–15 (2018)
21. C. Molnar. *Interpretable Machine Learning - A Guide for Making Black Box Models Explainable*. [online] Available: https://christophm.github.io/interpretable-ml-book/., 2019
22. T. Hastie, R. Tibshirani, M. Wainwright, *Statistical Learning with Sparsity* (CRC Press, The Lasso and Its Generalizations, 2015)
23. S. Boyd, L. Vandenberghe, *Convex Optimization* (Cambridge Univ. Press, Cambridge, UK, 2004)

Glossary

k-means The *k*-means algorithm is a hard clustering method. It aims at assigning data points to clusters such that they have minimum average distance from the cluster centre.

activation function Each artificial neuron within an ANN consists of an activation function that maps the inputs of the neuron to a single output value. In general, activation function is a non-linear map of the weighted sum of neuron inputs (this weighted sum is the activation of the neuron).

artificial intelligence Artificial intelligence aims at developing systems that behave rational in the sense of maximizing a long-term reward.

artificial neural network artificial neural artificial neural network is a graphical (signal-flow) representation of a map from features of a data point at its input to a predicted label at its output.

bagging bagging (or "bootstrap aggregation") is a generic technique to improve or robustify a given ML method. The idea is to use the bootstrap to generate perturbed copy of a given training set and then apply the original ML method to learn a separate hypothesis for each perturbed copy of the training set. The resulting set of hypotheses is then used to predict the label of a data point by combining or aggregating the individual predictions of each hypothesis. For numeric label values (regression) this aggregation could be obtained by the average of individual predictions.

Bayes estimator A hypothesis whose Bayes risk is minimal [1].

Bayes risk The Bayes risk of a given (fixed) hypothesis is the expectation of its loss incurred on (the realizations of) a random data point [1].

classifier A classifier is a hypothesis $h(\mathbf{x})$ that is used to predict a finite-valued label. Strictly speaking, a classifier is a hypothesis $h(\mathbf{x})$ that can take only a finite number of different values. However, we are sometimes sloppy and use the term classifier also for a hypothesis that delivers a real number which is then used in a simple thresholding to determine the predicted label value. For classifier, text, in a binary classification problem with label values $y \in \{-1, 1\}$, we refer to a

A. Jung, *Machine Learning*, Machine Learning: Foundations, Methodologies, and Applications, https://doi.org/10.1007/978-981-16-8193-6

linear hypothesis $h(\mathbf{x}) = \mathbf{w}^T \mathbf{x}$ as classifier if it is used to predict the label value according to $\hat{y} = 1$ when $h(\mathbf{x}) \geq 0$ and $\hat{y} = -1$ otherwise.

clustering   Clustering refers to the problem ot determining for each data point within clustering, text dataset to which cluster it belongs to. A cluster can be defined and represented in various ways, e.g., using representative data points ("cluster means") or an entire probability distribution (see GMM).

condition number   The condition number $\kappa(\mathbf{Q})$ of a psd matrix $\mathbf{Q}$ is the ratio of largest to smallest eigenvalue of a psd matrix $\mathbf{Q}$.

convex   A set C in $\mathbb{R}^n$ is called convex if it contains the line segment between any two points of that set. A function is called convex if its epigraph is a convex set [2].

data   A set of data points.

data augmentation   Data augmentation methods add synthetic data points to an existing set of data points. These synthetic data points might be obtained by perturbations (adding noise) or transformations (rotations of images) of the original data points.

data point   A data point is any object that conveys information [3]. Data points might be students, radio signals, trees, forests, images, RVs, real numbers or proteins. We characterize data points using two fundamentally different groups of properties. One group of properties is referred to as features and can be measured or computed in an automated fashion. Another group of properties is referred to as labels. The label of a data point represents a higher-level facts or quantities of interest. In contrast to features, determining the data point of a data point typically requires human experts (domain experts). Roughly speaking, ML is the study and design of methods for predicting the label of a data point based solely on its features.

dataset   With a slight abuse of notation we use the terms "dataset" or "set of data points" to refer to an indexed list of data points $\mathbf{z}^{(1)}, \ldots,$. Thus, there is a first data point $\mathbf{z}^{(1)}$, a second data point $\mathbf{z}^{(2)}$ and so on. Strictly speaking a dataset is a list and not a set [4].

decision region   Consider a hypothesis map h that can only take values from a finite set $\mathcal{Y}$. We refer to the set of features $\mathbf{x} \in \mathcal{X}$ that result in the same output $h(\mathbf{x}) = a$ as a decision region of the hypothesis h.

decision tree   A decision tree is a flow-chart like representation of a hypothesis map h. More formally, a decision tree is a directed graph which reads in the feature vector $\mathbf{x}$ of a data point at its root node. The root node then forwards the data point to one of its children nodes based on some elementary test on the features $\mathbf{x}$. If the receiving children node is not a leaf node, i.e., it has itself children nodes, it decision tree another test. Based on the test result, the data point is further pushed to one of its neighbours. This testing and forwarding of the data point is repeated until the data point ends up in a leaf node (having no children nodes). The leaf nodes represent sets (decision regions) constituted by feature vectors $\mathbf{x}$ that are mapped to the same function value $h(\mathbf{x})$.

deep net   We refer to a an ANN with a large number of hidden layers as a deep ANN or deep net.

deep net The rectified linear unit or "ReLU" is a popular choice for the activation function of a neuron within an ANN. It is defined as $g(z) = \max\{0, z\}$ with z being the weighted input of the neuron.

effective dimension The effective dimension $d_{\text{eff}}(\mathcal{H})$ of an infinite hypothesis space $\mathcal{H}$ is a measure of its size. Roughly speaking the effective dimension is equal to the number of independent tunable parameters or weights of the model. These parameters might be the weights used by linear map or the weights and bias terms of a ANN.

eigenvalue We refer to a number $\lambda \in \mathbb{R}$ as eigenvalue of a square matrix $\mathbf{A} \in \mathbb{R}^{n \times n}$ if there is a non-zero vector $\mathbf{x} \in \mathbb{R}^n \setminus \{\mathbf{0}\}$ such that $\mathbf{A}\mathbf{x} = \lambda\mathbf{x}$.

eigenvalue decomposition The task of computing the eigenvalues and corresponding eigenvectors of a matrix.

eigenvector An eigenvector of a matrix $\mathbf{A}$ is a non-zero,text vector $\mathbf{x} \in \mathbb{R}^n \setminus \{\mathbf{0}\}$ such that $\mathbf{A}\mathbf{x} = \lambda\mathbf{x}$ with some eigenvalue λ.

empirical risk The empirical risk of a given hypothesis on a given set of datapoints is the average loss of the hypothesis computed over all datapoints in that set.

empirical risk minimization Empirical risk minimization is the optimization problem of empirical risk the hypothesis with minimum average loss (empirical risk) on a given set of data points (the training set). Many ML methods are special cases of empirical risk minimization.

Euclidean space The Euclidean space $\mathbb{R}^n$ of dimension n refers to the space of all vectors $\mathbf{x} = (x_1, \ldots, x_n)$, with real-valued entries $x_1, \ldots, x_n \in \mathbb{R}$, whose geometry is defined by the inner product $\mathbf{x}^T\mathbf{x}' = \sum_{j=1}^{n} x_j x'_j$ between any two vectors $\mathbf{x}, \mathbf{x}' \in \mathbb{R}^n$ [9].

expectation maximization Expectation maximization is generic technique for estimating the parameters of a probabilistic model from data [10–12]. In general, this technique delivers an approximation to the maximum likelihood estimate for the model parameters.

explainable empirical risk minimization An instance of structural risk minimization that adds a regularization term to the training error in ERM. The regularization term is chosen to favour hypotheses that are intrinsically explainable for a user.

explainable machine learning Explainable ML methods aim at complementing predictions with explanations for how the prediction has been obtained.

feature map A map that transforms some raw features into a new feature vector. The new feature vector might be preferable over the raw features for several reasons. It might be possible to use linear hypothesis with the new feature vectors. Another reason could be that the new feature vector is much shorter and therefore avoids overfitting or can be used for a scatterplot.

feature space The feature space of a given ML application or method is constituted by all potential values that the feature vector of a data point can take on. Within this book the most frequently used choice for the feature space is the Euclidean space $\mathbb{R}^n$ with dimension n being the number of individual features of a data point.

features   Those properties of a data point that can be measured or computed in an automated fashion. For example, if a data point is a bitmap image, then we could use the red-green-blue intensities of its pixels as features. Features are sometimes referred to as "variables", "attributes" or "predictors" []. However, this book uses the term predictor in a different sense, i.e., as a hypothesis map used to predict a numeric label.

federated learning (FL)   Federated learning is an umbrella term for ML methods that train models in a collaborative fashion using decentralized data and computation.

Finnish Meteorological Institute   The Finnish Meteorological Institute is a government agency responsible for gathering and reporting weather data in Finland.

Gaussian mixture model   Gaussian mixture models (GMM) are a family of probabilistic models for data points. Within a GMM, the feature vector $\mathbf{x}$ of a data point is interpreted as being drawn from one out of k different multivariate normal (Gaussian) distributions indexed by $c = 1, \ldots, k$. The probability that the feature vector $\mathbf{x}$ is drawn from the c-th Gaussian distribution is denoted p_c. The Gaussian Mixture is parametrized by the probability p_c of $\mathbf{x}$ being drawn from the c-th Gaussian distribution as well as the mean vectors $\boldsymbol{\mu}^{(c)}$ and covariance matrices $\boldsymbol{\Sigma}^{(c)}$ for $c = 1, \ldots, k$.

gradient descent   Gradient descent is an iterative method for finding the minimum of differentiable function $f(w)$.

Hilbert space   A Hilbert space is a linear vector space that is equipped with an inner product between pairs of vectors. One important example for a Hilbert spaces is the Euclidean spaces $\mathbb{R}^n$, for some dimension n, which consists of Euclidean vectors $\mathbf{u} = \left(u_1, \ldots, u_n\right)^T$ along with the inner product $\mathbf{u}^T \mathbf{v}$.

Huber loss   The Huber loss is a mixture of the squared error loss and the absolute value of the prediction error.

hypothesis   A map (or function) $h : \mathcal{X} \to \mathcal{Y}$ from the feature space $\mathcal{X}$ to the label space $\mathcal{Y}$. Given a data point with features $\mathbf{x}$ we use a hypothesis map to estimate (or approximate) the label y using the predicted label $\hat{y} = h(\mathbf{x})$. ML is about learning (finding) a hypothesis map such that $\hat{y} \approx h(\mathbf{x})$ for any data point.

hypothesis space   Every practical ML method uses a specific hypothesis space, which we typically denote by $\mathcal{H}$. The hypothesis space of a ML method is a subset of all possible maps from the feature space to label space. The choice for the hypothesis space should take into account available computational infrastructure of statistical aspects. If the computational infrastructure allows for efficient matrix operations hypothesis space we expect a linear relation between feature values and label, a good first candidate for the hypothesis space is the space of linear maps.

i.i.d.   independent and identically distributed; e.g., "x, y, z are i.i.d. RVs" means that the joint probability distribution $p(x, y, z)$ of the RVs x, y, z factors into the product $p(x)p(y)p(z)$ of the marginal probability distributions of the variables x, y, z which are identical.

i.i.d. assumption   The i.i.d. assumption interprets data points of a dataset as the realizations of i.i.d. random variables.

label   A higher level fact or quantity of interest associated with a data point. If a data point is an image, its label might be the fact that it shows a cat (or not).

label space   Consider a ML application that involves data points characterized by features and labels. The label space of a given ML application or method is constituted by all potential values that the label of a data point can take on. A popular choice for the label space in regression problems (or methods) is $\mathcal{Y} = \mathbb{R}$. Binary classification problems (or methods) use label spaces that consist of two different elements, e.g., $\mathcal{Y} = \{-1, 1\}$, $\mathcal{Y} = \{0, 1\}$ or $\mathcal{Y} = \{\text{``cat image''}, \text{``no cat image''}\}$

law of large numbers   The law of large numbers refers to the convergence of the partial sums of i.i.d. RVs to the (common) expectation these RVs.

learning rate   Consider an iterative method for finding or learning a good choice for a hypothesis. Such an iterative method repeats similar computational (update) steps that adjust or modify the current choice for the hypothesis to obtain an improved hypothesis. A prime example for such an iterative learning method is GD and its variants (see 5). We refer by learning rate to any parameter of an iterative learning method that controls the extent by which the current hypothesis might be modified or improved in each iteration. A prime example for such a parameter is the step size used GD. Within this book we use the term learning rate mostly as a synonym for the step size of (a variant of) GD.

least absolute deviation regression   Least absolute deviation regression uses the average of the absolute precondition errors to find a linear hypothesis.

linear classifier   A classifier $h(\mathbf{x})$ maps the feature vector $\mathbf{x} \in \mathbb{R}^n$ of a data point to a predicted label $\hat{y} \in \mathcal{Y}$ out of a finite set of label values $\mathcal{Y}$. We can characterize such a classifier equivalently by the decision regions $\mathcal{R}^{((a))} := \{\mathbf{x} \in \mathbb{R}^n : \hat{y} = (a)\}$, for every possible label value $a \in \mathcal{Y}$. Linear classifiers are such that the boundaries between the regions $\mathcal{R}^{(a)}$ are hyperplanes in $\mathbb{R}^n$.

linear regression   Linear regression aims at learning a linear regression map to predict a numeric label based on numeric features of a data point. The quality of a linear hypothesis map is typically measured using the average squared error loss incurred on a set of labeled data points (the training set).

logistic loss   Consider a data point that is characterized by the features $\mathbf{x}$ and logistic loss binary label $y \in \{-1, 1\}$. We use a hypothesis h to predict the label y solely from the features $\mathbf{x}$. The logistic loss incurred by a specific hypothesis h is defined as (2.12).

logistic regression   Logistic regression aims at learning a linear hypothesis map to predict a binary label based on numeric features of a data point. The logistic regression of a linear hypothesis map (classifier) is measured using its average logistic loss on some labeled data points (the training set).

loss   With a slight abuse of language, we use the term loss either for loss function itself or for its value for a specific pair of data point and hypothesis.

loss function   A loss function is a map

$$L : \mathcal{X} \times \mathcal{Y} \times \mathcal{H} \to \mathbb{R}_+ : \big((\mathbf{x}, y), h\big) \mapsto L((\mathbf{x}, y), h)$$

which assigns a pair consisting of a datapoint, with features $\mathbf{x}$ and label y, and a hypothesis $h \in \mathcal{H}$ the non-negative real number $L((\mathbf{x}, y), h)$. The loss value $L((\mathbf{x}, y), h)$ quantifies the discrepancy between the true label y and the predicted label $h(\mathbf{x})$. Smaller (closer to zero) values $L((\mathbf{x}, y), h)$ mean a smaller discrepancy between predicted label and true label of a data point. Figure 2.11 depicts a loss function for a given data point, with features $\mathbf{x}$ and label y, as a function of the hypothesis $h \in \mathcal{H}$.

maximum   Given a set of real numbers, the maximum is the largest of those numbers.

mean   The expectation of a real-valued random variable.

metric   A metric refers to a loss function that used solely for the final performance evaluation of a learnt hypothesis. The metric is typically a loss function that allows for easy metric, text (such as the 0/1 loss (2.9)) but is not suitable for being used within ERM to learn a hypothesis. For learning a hypothesis via ERM, we typically prefer loss functions that depend smoothly on the (parameters of the) hypothesis. Examples for such smooth loss functions are the squared error loss (2.8) and the logistic loss (2.12).

minimum   Given a set of real numbers, the minimum is the smallest of those numbers.

model   We use the term model as a synonym for hypothesis space

multi-label classification   Multi-label classification problems and methods involve data points that are characterized by several individual labels.

non i.i.d. data   A dataset that cannot be well modelled as realizations of i.i.d. RVs.

non-i.i.d.   See non-i.i.d. data.

nonsmooth   We refer to a function as non-smooth if is not smooth [5].

outlier   Many ML methods are motivated by the i.i.d. assumption which interprets data points as i.i.d. realizations of RVs with a common probability distribution. The i.i.d. assumption is typically useful when the statistical properties of the data generation process are stationary (time-invariant). In some applications the data might consists of a majority of "regular" data points that conform with an i.i.d. assumption and a small number of data points that have fundamentally different statistical properties compared to the bulk of regular data points. We refer to a data point that substantially deviates from the statistical properties of the majority of data points as an outlier.,

parameters   The parameters of ML model are tunable (learnable or adjustable) quantities that allow to choose between different hypothesis maps. For example, the linear model $\mathcal{H} := \{h : h(x) = w_1 x + w_2\}$ consists of all hypothesis maps $h(x) = w_1 x + w_2$ with a particular choice for the parameters w_1, w_2. Another example of parameters are the weights assigned to the connections of an artificial neural network.

positive semi-definite   A symmetric matrix $\mathbf{Q} = \mathbf{Q}^T \in \mathbb{R}^{n \times n}$ is referred to as positive semi-definite if $\mathbf{x}^T \mathbf{Q} \mathbf{x} \geq 0$ for every vector $\mathbf{x} \in \mathbb{R}^n$.

predictor   We refer to a hypothesis whose function values are real numbers as a predictor. Given a data point with features $\mathbf{x}$, the predictor value

$h(\mathbf{x}) \in \mathbb{R}$ is used as a prediction (estimate/guess/approximation) for the true numeric label $y \in \mathbb{R}$ of the data point.

principal component analysis   The principal component analysis determines a given number of new features that are obtained by a linear transformation (map) of the raw features.

probabilistic principal component analysis   Probabilistic principal component analysis combines PCA with a probabilistic model for data. This probabilistic model allows to interpret the goal of PCA as the estimation of the parameters of an underlying probabilistic model for the data generation process.

random forest   A random forest is an ensemble of decision trees, each one learnt or fitted to a perturbed copy of the original dataset.

random variable   Formally, a random variable is a map from a set of elementary events to a set of values. The set of elementary events is equipped with a probability measure. A real-valued random variable maps elementary events to real numbers $\mathbb{R}$. A discrete random variable maps elementary events to a finite set such as $\{-1, 1\}$ or {cat, no cat}. A vector-valued random variable maps elementary events to the Euclidean space $\mathbb{R}^n$ with some dimension $n \in \mathbb{N}$.

regularization   The term regularization refers to different techniques for modifying ERM to learn a hypothesis that performs well also outside the training set used in ERM. One specific approach to regularization is by adding a penalty or regularization term to the ERM objective function (which is the average loss on the training set).

sample   A finite sequence (list) of data points $\mathbf{z}^{(1)}, \ldots, \mathbf{z}^{(i)}$ that is obtained or interpreted as the realizations of m i.i.d. RVs with the common probability distribution $p(\mathbf{z})$. The length m of the sequence is referred to as the sample size.

sample size   The number of individual data points contained in a dataset that is obtained from realizations of i.i.d. RVs.

scatterplot   A visualization technique that depicts data points by markers in a two-dimensional plane.

semi-supervised learning   Semi-supervised learning methods use (large amounts of) unlabeled data points to support the learning of a hypothesis from (a small number of) labeled data points [6].

smooth   We refer to a real-valued function as smooth if it is differentiable and its gradient is continuous [5, 7].

soft clustering   Soft clustering refers to the problem ot determining for each data point within a dataset, the degree of belonging to a particular cluster.

stochastic gradient descent (SGD)   Stochastic gradient descent is obtained from GD by replacing the gradient of the objective function by a noisy (or stochastic) estimate.

structural risk minimization   Structural risk minimization is the problem of finding the hypothesis that optimally balances the average loss (empirical risk) on a training set with a structural risk term. The regularization term penalizes a hypothesis that is not robust against (small) perturbations of the data points in the training set.

support vector machine   A ML method that learns a linear map such that the classes are maximally separated in the feature space ("maximum margin"). This maximum margin condition is equivalent to minimizing a regularized variant of the hinge loss (2.11).

training error   Consider a ML method that aims at learning a hypothesis $h \in \mathcal{H}$ out of a hypothesis space. We refer to the average loss or empirical risk of a hypothesis $h \in \mathcal{H}$ on a dataset as training error if it is used to choose between different hypotheses. The principle of ERM is find the hypothesis $h^* \in \mathcal{H}$ with training error training error. Overloading the notation a bit, we might refer by training error also to the minimum empirical risk achieved by the optimal hypothesis $h^* \in \mathcal{H}$.

training set   A set of data points that is used in ERM to train a hypothesis $\hat{h}$. The average loss training set $\hat{h}$ on the training set is referred to as the training error. The comparison between training and validation error informs adaptations of the ML method (such as using a different hypothesis space).

validation error   Consider a hypothesis $\hat{h}$ which is obtained by ERM on a training set. The average validation error of $\hat{h}$ on a validation set which is different from the training set, is referred to as the validation error.

validation set   A set of data points that has not been used as training set in ERM to train a hypothesis $\hat{h}$. The average loss of $\hat{h}$ on the validation set is referred to as the validation error and used to diagnose the ML method. The comparison between training and validation error informs adaptations of the ML method (such as using a different hypothesis space).

Vapnik–Chervonenkis (VC) dimension   The VC dimension is maybe the most widely used concept for measuring the size of infinite hypothesis spaces. For a precise definition of the VC dimension and discussion of its applications in ML we refer to [8].

variance   The expectation $\mathbb{E}\{(x - \mathbb{E}\{x\})^2\}$ of the squared difference between a real-valued random variable and its expectation.

weights   We use the term weights synonymously for a finite set of parameters within a model. For example, the linear model consists of all linear maps $h(\mathbf{x}) = \mathbf{w}^T \mathbf{x}$ that read in a feature vector $\mathbf{x} = (x_1, \ldots, x_n)^T$ of a data point. Each specific linear map is characterized by specific choices for the parameters for weights $\mathbf{w} = (w_1, \ldots, w_n)^T$.

References

1. E.L. Lehmann, G. Casella, *Theory of Point Estimation*, 2nd edn. (Springer, New York, 1998)
2. S. Boyd, L. Vandenberghe, *Convex Optimization* (Cambridge University Press, Cambridge, UK, 2004)
3. T.M. Cover, J.A. Thomas, *Elements of Information Theory*, 2nd edn. (Wiley, Hoboken, NJ, 2006)
4. P. Halmos, *Naive Set Theory* (Springer, Berlin, 1974)
5. Y. Nesterov, *Introductory Lectures on Convex Optimization, Vol. 87 of Applied Optimization* (Kluwer Academic Publishers, Boston, MA, 2004). (A basic course)
6. O. Chapelle, B. Schölkopf, A. Zien (eds.), *Semi-Supervised Learning* (The MIT Press, Cambridge, MA, 2006)
7. S. Bubeck, Convex optimization: algorithms and complexity. Foundations and Trends in Machine Learning **8**(3–4), 231–357 (2015)
8. S. Shalev-Shwartz, S. Ben-David, *Understanding Machine Learning: From Theory to Algorithms* (Cambridge University Press, Cambridge, UK, 2014)
9. W. Rudin, *Principles of Mathematical Analysis*, 3rd edn. (McGraw-Hill, New York, 1976)
10. C.M. Bishop, *Pattern Recognition and Machine Learning* (Springer, Berlin, 2006)
11. T. Hastie, R. Tibshirani, J. Friedman, *The Elements of Statistical Learning* Springer Series in Statistics. (Springer, New York, 2001)
12. M.J. Wainwright, M.I. Jordan, *Graphical Models, Exponential Families, and Variational Inference*, Foundations and Trends in Machine Learning, vol. 1. (Now Publishers, Hanover, MA, 2008)

Index

A
Active learning, 84
Artificial intelligence, 10
Audio recording, 25
Autoencoder, 176

B
Bagging, 141
Baseline, 114
Bayesian network, 193
Benchmark, 114
Binary classification, 26
Black box, 190

C
City planner, 191
Cluster, 16, 24
Clustering, 13, 35
Confusion matrix, 46
Convex, 78, 86
Core node, 168

D
Data, vii
Data augmentation, 24, 141
Data engineering, 20
Dataset, 21
DBSCAN, 169
Decision boundary, 68
Deep learning, vii, 73
Deep net, 73

Dermatologist, 191
Differentiable, 38
Discriminative methods, 93

E
Effective dimension, 36
Empirical risk, 45
Euclidean space, 25
Expectation maximization, 165
Explainable ML, viii

F
Feature, 21
Feature learning, 13
Feature map, 35
Feature space, 23

G
Gaussian mixture model (GMM), 163, 166
Generalization, 32
Generative methods, 93
Gradient, 16
Gradient-based methods, 99
Groups, 24

H
High-dimensional data, 88
Hinge loss, 42
Huber loss, 60
Hypothesis space, vii37

I
Imbalanced data, 46
Imputation, 21
Information theory, 190
Inner product, 5
Iterative methods, 99

K
Kernel SVM, 66
k-means, 155

L
Label, vii, 26
Label space, 26
Learning rate schedule, 109
Learning task, 136
Linear classifier, 34
Linea regression, 57
Linearly independent, 116
Linear separable, 67
Linear span, 115
Loss, vii

M
Matrix inversion, 89
Maximum likelihood, 29
Mean, 29
Mean squared error, 59
Mean squared estimation error, 127
Missing data, 21
Model, vii
Model agnostic, 190
Multi-class classification, 27
Multi-label classification, 27

N
Naive Bayes, 93
Non-convex, 85
Nonsmooth, 43

O
Online learning, 94
Optimization, 6

P
Parameter, 2
Predict, 26
Prediction error, 49
privacy-preserving ML, viii
Probabilistic principal component analysis,
          181
Probability theory, 9
Pseudoinverse, 89

R
Random forest, 141
Rectified linear unit, 73
Regularization, 67, 135
Regularization parameter, 67, 138
Regularizer, 138
Reinforcement learning, 14
Reward, 12
Ridge regression, 139

S
Sample covariance matrix, 177
Sample size, 21
Scatterplot, 5
Semi-supervised learning, 136
Sigmoid function, 73
Smooth, 78, 86
Soft clustering, 92
Spectrogram, 25
Stochastic gradient descent, 109
Supervised ML, 12
Support vectors, 68

U
Underfitting, 32
Unsupervised ML, 13

V
Validation error, 133
Variance, 29

W
Weights, 2

Printed in the United States
by Baker & Taylor Publisher Services